Prosthetic Designs for Restoring Human Limb Function

William Craelius

Prosthetic Designs for Restoring Human Limb Function

 Springer

William Craelius
Department of Biomedical Engineering
Rutgers, The State University of New Jersey
Piscataway, NJ, USA

ISBN 978-3-030-31076-9   ISBN 978-3-030-31077-6 (eBook)
https://doi.org/10.1007/978-3-030-31077-6

© Springer Nature Switzerland AG 2022

This work is subject to copyright. All rights are reserved by the Publisher, whether the whole or part of the material is concerned, specifically the rights of translation, reprinting, reuse of illustrations, recitation, broadcasting, reproduction on microfilms or in any other physical way, and transmission or information storage and retrieval, electronic adaptation, computer software, or by similar or dissimilar methodology now known or hereafter developed.

The use of general descriptive names, registered names, trademarks, service marks, etc. in this publication does not imply, even in the absence of a specific statement, that such names are exempt from the relevant protective laws and regulations and therefore free for general use.

The publisher, the authors, and the editors are safe to assume that the advice and information in this book are believed to be true and accurate at the date of publication. Neither the publisher nor the authors or the editors give a warranty, expressed or implied, with respect to the material contained herein or for any errors or omissions that may have been made. The publisher remains neutral with regard to jurisdictional claims in published maps and institutional affiliations.

This Springer imprint is published by the registered company Springer Nature Switzerland AG
The registered company address is: Gewerbestrasse 11, 6330 Cham, Switzerland

Preview of the Book: Designing Smart Prostheses

As an engineer, your goal is to make things that perform better than existing products. Here, your goal is a better prosthesis: It should faithfully execute its user's commands, adapt to the dynamic environment, and be easy to produce, be durable, *et cetera*. This book teaches methods to achieve that goal. This book emphasizes the opportunities for Engineers to innovate and does not cover the very important techniques of fitting and installing prostheses, which is the job of Certified Prosthetists.

You, the designer of limb prostheses, must restore your clients' natural mobility to the highest degree possible. The prosthesis must faithfully execute its user's commands and adapt to various environments and tasks. To achieve this goal, the designer needs access to a multidisciplinary knowledge base, with segments of biomedical, mechanical, electrical, computer and materials engineering, as well as anatomy, physiology, and various subfields of medicine and surgery. Advancing the field also requires knowledge of its historical development and both its successes and failures. The book teaches from each of those disciplines, employing modeling techniques based on Matlab/Simulink for quicker understanding (*Octave* is an open-source alternative).

The rapid advancement in technologies in many fields has set the stage for a revolution in the art of restoring human mobility. The author's development of new methods for upper-limb prosthetic control is one such example and is being widely adapted in practice. Other examples include the magneto-hydraulic knee and the use of URL tags for control of mechanical hands. This book seeks to stimulate innovative thinking with a wide range of relevant disciplines.

There is a vast and growing body of knowledge from many disciplines that relate to prosthetic design. This text can only present a fraction of such knowledge, and rather emphasizes ideas and tools for creating innovative devices. As engineers, you will naturally want to design better prostheses and will practice doing so by practical modeling projects using Matlab and Simulink. These tools will allow you to design and build and test realistic models of prosthetic limbs. The exercises will encourage creativity and a better understanding of human motion and its restoration.

Designing Smart Prostheses

The smart prosthesis should replicate the function of the original limb as closely as possible. This requires not only appropriate hardware but an interface that can establish a seamless and open communication pathway (two-way) with its user. The interface should transmit two-way mechanical and biological signals between. The interface must be (1) compatible with the user's anatomy, physiology, and chemistry; it must support and transmit mechanical actions, and (2) communicate appropriately with biological signals directing the user's commands. The goal of modern prosthetic design is a "smart" interface that should possess three key characteristics: durability, seamlessness, and healthy communication with surrounding tissues.

Our goal in this book is to teach Engineers at least a representative fraction of the current state of knowledge and practice in the field and to highlight the areas most in need of their attention. This idea was articulated in *Science magazine* in 2002: "When humans replace a missing body part with an artificial one, they begin an intimate relationship with a partner they barely know". The key to success for such a relationship may be no different than that found in marriage manuals: communication. Unfortunately, for bionic parts, communication is the weakest link in the chain of components that includes electronics, computing, actuators, mechanisms, and materials, all of which are adequate for the application.

The final requirement of a smart interface is healthy physical communication with the tissues surrounding it. This two-way communication is necessary for the interface to adapt to the ever-changing environment of the body—remodeling, self-healing, and growing as biological conditions change. This type of behavior wherein an artificial material works together with a biological material is called "biopermissive." The biopermissive interface allows bidirectional biological signaling (mechanical, chemical, and electrical) and uses the body's nutrients, resources, and waste disposal mechanisms, for both metabolic waste and mechanical wear. This also includes communicating with the human sensory system for restoration of tactile and proprioceptive sensation. The smart interface must enable the prosthetic mechanics to work with the body as a living part of the system; this remains the primary challenge for prosthetic designs.

While it is within the Engineer's job description to innovate, he/she must avoid adding unnecessary complexity to products that already work pretty well. The design needs to consider (1) what degree of hardware intelligence (processing power) is needed for useful function?, (2) how much technology can be incorporated without making the device unfriendly to the user?, and (3) how can the prosthesis adapt to the user?

The Professions of Prosthetics and Orthotics

The practice of Prosthetics and Orthotics exploits the technologies of imaging, modeling, 3D printing, artificial intelligence, and robotics, as described in this text. The *Prosthetist* fabricates and installs replacement limbs to persons following amputation, and the *Orthotist* builds a device to assist a weakened or vulnerable limb. The professions of *Prosthetist* and *Orthotist* formally began in the 1950s with the introduction of courses at UCLA, NYU, and Northwestern University. In 1998, Rutgers University inaugurated the first M.S. program in Prosthetics and served as a model for subsequent programs, which now require an M.S. degree for certification.

The *Orthotist* creates devices that provide support to parts of a patient's body to compensate for paralyzed muscles, provide relief from pain, or prevent orthopedic deformities from progressing. The Orthotist must understand the anatomy, physiology, and pathology of the human body and the forces that may be applied to the body to enhance or maintain a patient's quality of life. Both prosthetists and orthotists need skills to control the interaction between devices and the individual patient. They must also understand the properties of the materials and components from which the devices are made in order to ensure safe and effective clinical practice.

Because each patient and his or her amputation are unique, each prosthetic limb must be custom fitted and then fabricated by the Prosthetist. Because prosthetists work to interface artificial devices with the human body, they need a wide range of skills in areas such as engineering, anatomy, and physiology. In 1949, upon the recommendation of the Association, the *American Board for Certification in Orthotics and Prosthetics*, Inc. was established to ensure that prosthetists and orthotists met certain standards of excellence, much in the manner that certain physicians' specialty associations are conducted. Examinations are held annually for those desiring to be certified. In addition to certifying individuals as being qualified to practice, the *American Board for Certification* approves individual shops, or facilities, as being satisfactory to serve the needs of amputees and other categories of the disabled requiring mechanical aids.

Prior to 1957, medical schools offered little in the way of training in prosthetics to doctors and therapists. To encourage the inclusion of prosthetics into medical and paramedical curricula, the National Academy of Sciences organized the Committee on Prosthetics Education and Information, and as a result of the efforts of this group many schools have adopted courses in prosthetics at both undergraduate and graduate levels.

How to Use the Book

Much of the readings and graphics are from published articles, conferences, and books, as cited. Your learning should be supplemented by reliable sources from the web. When doing your homework and projects, you should search the appropriate literature as needed. Please remember that website data, especially, represent *potential* knowledge but are not necessarily accurate. Any source used should be cited, but should not be reproduced in your answers without quotes.

Learning is "problem-based," and for homework, the problems will usually be multi-step, that is, you will need to analyze a real situation and perform more than one calculation step to arrive at the final answer. For example, "specify battery size for a robotic hand." This would involve finding real data on typical hand usage, estimating duty cycle and motor power for an artificial hand, and then finding energy need. One of the tools will be Simulink, solving differential equations in a graphical environment. Sample data will be on the Sakai site for your analysis. The appendices contain formulae, equations, and parameters for reference in modeling.

This text emphasizes the application of engineering principles to prosthetic design. It does not cover the "nitty-gritty" of prosthetic restoration, which is a job of trained professionals.

Motivation: A prosthesis is a device that replaces a body part, i.e., an artificial limb or sensory organ. The field of Prosthetics is multidisciplinary, with contributions from biomedical, mechanical, electrical and materials engineering, as well as anatomy, physiology, and various subfields of medicine, and surgery. You, the Engineer, are needed to design better replacements for limbs. This book presents the basics of prosthetic technologies and their future possibilities. The following quotes from 1955 summarizes the importance of engineers' involvement in this field:

> At the end of World War II, our knowledge of human locomotion was still quite incomplete, and such knowledge as existed was only poorly understood. Thus it was that, when approached in September of 1945 by the then Committee on Artificial Limbs of the National Research Council, the representatives of the College of Engineering and of the Medical School of the University of California could point to the necessity of the adoption of a long-term outlook which envisioned the study of the fundamentals of human locomotion, of the amputee who must wear a lower-extremity prosthesis, and of the prosthesis itself. It could be shown that the experience of 400 years in trial-and-error techniques had offered little and that a firm basis for progress could be established only by a systematic approach. It was predicted that at least 7 years of study would be required to collect the fundamental data necessary for improved design of artificial legs [1].

Reference

1. Saunders JBdM. Prelude, prophecy, and promise. O&P Library. Artif Limbs. 1955;1(2): 1–3.

Contents

1 Introduction 1
   1.1   Restoring Limbs 1
   1.2   The Need 1
   1.3   Highlights of Prosthetic Development 2
   1.4   Prosthetic Advances 3
        1.4.1   Lower Limb 3
        1.4.2   Upper Limbs 5
        1.4.3   Surgical Techniques 6
        1.4.4   Causes of Limb Loss 7
   1.5   The Costs of Amputation 9
   1.6   Challenges for Lower-Limb Amputees 9
   1.7   Types of Lower-Limb (LL) Amputations 10
        1.7.1   Below-Knee (BK) Amputations 10
        1.7.2   Knee Disarticulation 11
        1.7.3   Above-Knee Amputations 11
        1.7.4   Hip Disarticulation and Hemipelvectomy 11
   1.8   Upper-Extremity Amputations 12
        1.8.1   Functional Comparison of Upper and Lower Limbs ... 12
        1.8.2   Partial-Hand Amputations 12
        1.8.3   Wrist Disarticulation 13
        1.8.4   Below Elbow 13
        1.8.5   Above Elbow 13
        1.8.6   Shoulder Disarticulation and Forequarter Amputation .. 13
   1.9   The Postsurgical Period 14
   1.10 Regulation of Prosthetic and Orthotic (P&O) Devices 15
   1.11 Multidisciplinary Approach to Prosthetic Design 15
   1.12 Exercises 16
   References 16

2	**Human Limb Biomechanics**.		17
	2.1	Introduction: Divisions of Biomechanics.	17
	2.2	Forces Within Living Limbs.	18
	2.3	Stress on Limbs	18
	2.4	Free Body Diagrams.	19
		2.4.1 Balance of Forces and Torques.	20
	2.5	Kinematics and Kinetics.	21
	2.6	Dynamic Analysis.	23
	2.7	Uniaxial Stress and Strain	25
	2.8	Beam Analysis of Structures	26
	2.9	Beam Buckling.	27
	2.10	Applied Stress to Cube.	27
	2.11	Exercises.	29
	Reference		30
3	**Metrics and Mechanics of Human Limbs**.		31
	3.1	Introduction	31
	3.2	Anthromorphometry.	31
	3.3	Measuring Body Segments.	35
	3.4	Anatomical Coordinates.	36
	3.5	Measuring Human Movement and Dynamics	36
	3.6	Human Walking Kinetics	38
		3.6.1 Modern Methods of Tracking Human Motion	40
		3.6.2 Measuring Joint Dynamics.	40
		3.6.3 Muscle Activities During Gait	44
		3.6.4 Measuring Locomotion.	45
	3.7	Exercises.	50
4	**Lower-Limb Prostheses**.		51
	4.1	Introduction	51
	4.2	Leg Functional Anatomy	52
		4.2.1 Knee Function After BK Amputation	52
	4.3	Natural and Prosthetic Knee Motion During a Gait Cycle.	56
	4.4	Prostheses for the AK Case	58
	4.5	Mimicking the Human Lower Limb.	60
		4.5.1 Gait Muscles Summarized	63
	4.6	Hip Disarticulation and Bilateral Leg Cases	66
		4.6.1 Prostheses for Hip Disarticulation and Hemi-pelvectomy Cases.	66
	4.7	Restoring Ambulation.	67
	4.8	Exercises.	68
	Bibliography.		68
5	**Upper Limb Function**.		71
	5.1	Introduction	71
	5.2	Human Hand Function	72

	5.3	Functional Anatomy, Motions, and Dynamics of the UL	72
	5.4	The Hand	72
	5.5	Common Upper-limb Positions and Forces	73
	5.6	Hand Control	78
	5.7	Manipulation	80
	5.8	Arm Actuators	80
	5.9	UL Joint Torques	81
	5.10	Hand Synergies	82
	5.11	Exercises	83
6	**The Human-Machine Interface**		**85**
	6.1	Restoring UL Function	85
	6.2	HMIs	85
	6.3	Example of an Advanced Internally Powered Prosthesis	87
	6.4	Externally Powered Prostheses (EPPs)	88
	6.5	Increasing Degrees of Freedom in the Upper Limb Prosthesis	89
		6.5.1 Myoelectrical Signals as HMI Input	89
	6.6	Characteristics of MYOE	91
	6.7	Radio-Transmitting Electrodes Implants	92
	6.8	Targeted Muscle Reinnervation	92
	6.9	Vectorializing MYOE Signals	93
	6.10	Direct Brain Control	97
	6.11	Alternative HMIs	98
	6.12	Comparing FMG with EMG	99
	6.13	Developing a FMG-Based HMI	102
		6.13.1 Silicon Sleeve with Integrated FSR Sensors	104
	6.14	Biomimetic Control	107
	6.15	Volitional Signal Processing	108
	6.16	Prosthetic Developments	110
	6.17	Sensory Receptors and Local Feedback	111
	6.18	Adaptive Neuromotor Learning	112
	6.19	Cosmetic Restoration and Agency	114
	References		114
7	**Prosthetic Control Systems**		**117**
	7.1	Introduction	117
	7.2	Artificial Control	117
	7.3	Natural Control of Limbs	118
	7.4	Prosthetic Control	122
	7.5	Adaptability of the Neuromotor Control System: Gravity Compensation	122
	7.6	Control Logic of Limb Function	124
	7.7	Gait: Controlled Falling	126
	7.8	Walking Coordination	126
	7.9	Tactile Feedback for Control	129
	7.10	Exercises	130

8 Limb-Prosthetic Interface ... 133
- 8.1 Introduction ... 133
- 8.2 The Socket Environment ... 133
- 8.3 LL Socket Fitting ... 134
- 8.4 Upper-Limb Sockets ... 134
- 8.5 Fitting the BK Residuum ... 135
- 8.6 Ground-Reaction Forces ... 137
- 8.7 Installing the Leg Prosthesis ... 139
- 8.8 Socket-Skin Interface ... 142
- 8.9 Forces on the Skin During Ambulation ... 143
- 8.10 Materials for Sockets ... 145
- 8.11 Protecting Tissue in the Socket ... 145
- 8.12 Material Selection for Structures ... 146
- 8.13 Laminates ... 146
- 8.14 Material Properties of Selected Prosthetic Laminates ... 147
- 8.15 Dynamic Strengths of Laminates ... 149
- 8.16 Preserving the Residuum ... 149
- 8.17 Custom Designing and Fitting Limb Sockets ... 150
- 8.18 Summary ... 152
- 8.19 Endoskeletal ... 152
- 8.20 Exercises ... 153
- Bibliography ... 153

9 Energetics of Ambulation ... 155
- 9.1 Introduction ... 155
- 9.2 Natural Energy Sources for Ambulation ... 155
- 9.3 Cost of Transport (COT) ... 156
- 9.4 Artificial Energy Sources for Ambulation ... 156
- 9.5 Measuring Energy of Gait ... 157
- 9.6 Energetic Contributors to COT ... 158
- 9.7 Ambulatory Efficiency ... 160
- 9.8 Energy-Efficient Walking Speeds ... 161
  - 9.8.1 Walking, Power, and Work Units ... 161
- 9.9 Walking Energy Harvesting ... 162
- 9.10 Comparative Energetics ... 164
- 9.11 Testing with Amputees ... 167
- 9.12 Problems ... 167

10 Advancing Prosthetic Designs ... 169
- 10.1 Trends in Prosthetic Technology ... 169
- 10.2 Focusing on Users ... 170
  - 10.2.1 Improving Manipulation ... 170
  - 10.2.2 Sensory Feedback for Dexterity ... 170
  - 10.2.3 Direct Brain Control ... 173
- 10.3 Engaging Users in UL Prosthetic Design ... 173
  - 10.3.1 Reasons for Abandonment of Prosthesis ... 175

10.4	Optimal Restoration of Hand Function	176
	10.4.1 Agency as a Factor in HMI Performance	178
10.5	Restoring Agency	180
10.6	Developing Better HMIs	180
	10.6.1 Hardware Needs	180
10.7	Exploiting Adaptability	181
10.8	Agency as a Key to Controlling the Prosthesis	181
	10.8.1 Matching Prosthesis to User	182
10.9	Conclusion	184
10.10	Exercises	184
Bibliography		185

Appendix: Computational Modeling 189

Index 219

Chapter 1
Introduction

Goal: To learn the historical developments of prostheses and major causes for amputation.

1.1 Restoring Limbs

Restoring the natural functions of a lost limb requires understanding of its anatomy, function, connections, and control. These attributes need to be translated into engineering specifications of materials, shape, mechanics, power requirements, and performance. All these features should work together to restore natural functionality, including tactile and proprioceptive feedback normally provided by sensory organs in natural limbs, while preserving the tissue integrity of the residuum. Since every individual is different, each prosthesis is customized to varying degrees. While it is unlikely that an artificial limb, however elaborate, can ever completely substitute for a natural member, the rapid advances in prosthetic-related technology promise ever more realistic complete restorations.

1.2 The Need

The population of persons living with amputations is expected to rise to 3.6 million persons by the year 2050, and the need for advanced prosthetic devices is thus expected to grow proportionally. The main cause of amputations in the U.S. today is dysvascularity, a complication that is usually secondary to diabetes or related disease. There is an estimated 30 million people, roughly split between male and female, living with diabetes in the U.S. currently [1] (2019). Despite these large and growing figures, the target for prosthetic devices remains small relative to consumer goods but is nevertheless an extremely important one, requiring maximal exploitation of technology.

1.3 Highlights of Prosthetic Development

Until recent times, artificial limbs were made by artisans without formal education, working with primitive materials and methods. It is instructive to look at their innovations, since some of today's technology can be traced directly back to ancient times. Ancient pottery from about 300 B.C. depicts prostheses constructed of copper and wood (Fig. 1.2), which are displayed at the *Cathedral of Lescar*, France. The prostheses appear to be wooden sticks attached to the residual limb (Fig. 1.1 on the pot), which served as leg substitutes for centuries. As proof, the Greek historian Herodotus wrote in *"History"* (484 B.C.) of a Persian soldier who replaced his severed foot with a wooden version.

The *Alt-Ruppin* hand from the fifteenth century was displayed at museums in Florence, Italy (Fig. 1.2). Mechanical hands with almost the same functionality as today's hands were invented by innovators such as Knight Gotz von Berlichingen in the 1500s [2].

The use of ligatures for limb salvage was taught by Hippocrates, and his followers up to the present have used it successfully. More restorative surgical procedures have been developed, and surgeons now have a much better set of surgical tools, imaging, and prosthetic hardware with which to maximize survival and functional restoration.

Examples of early technological improvements in limb salvage and restoration are the tourniquet by Morel in 1674, better surgical tools and procedures, including metallic prostheses with articulating joints for above-knee amputations (Figs. 1.3 and 1.4). The surgeon, Verduin, improved the knee joint, resembling the thigh-corset below-knee limb still in use today (Fig. 1.3) [3].

In 1800, Potts introduced one of the first functional leg prostheses that allowed for foot articulation, which greatly improved ambulation. The artificial foot was

Fig. 1.1 Evidence of the existence of prosthetics possibly as early as 486 A.D. These are artistic representations of discovered antiques both depicting amputees with a wooden replacement below the knee [2]

1.4 Prosthetic Advances

Fig. 1.2 Example of the Alt-Ruppin hand developed around the 1400s. This prosthetic hand includes advancements such as a hinged wrist, and two sets of movable fingers controlled by buttons on the palm. Putti, V., *Chir. d. org. di movimento*, 1924–1925 [2]

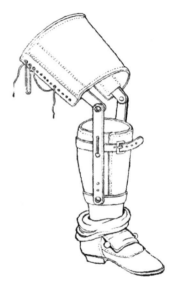

Fig. 1.3 Verduin Leg, an artificial limb with a hinged knee allowing for movement of the knee joint (1696) [2]

connected to the wooden shank by artificial tendons, simulating to a degree the natural leg. This apparatus allowed coordination of toe lift with knee flexion. The Potts leg underwent many refinements, which were incorporated by other limb fitters, including Americans, and became known as the "American leg."

1.4 Prosthetic Advances

1.4.1 Lower Limb

The U.S. Civil War produced >30,000 amputees, a tragedy that was addressed by the U.S. government, which supported artificial limb replacements for the Veterans. One of these Veterans, J. E. Hanger, lost a leg in the Civil War, and he was inspired to re-design the so-called American leg by inserting rubber bumpers in the wooden foot to simulate the cushioning and flexibility provided by ankle joint. This feature

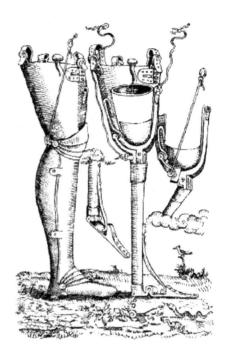

Fig. 1.4 Early prosthesis with articulating joints (middle sixteenth century). From Pare, A., Oeuvres Completes, Paris, 1840. From the copy in the *National Library of Medicine*

is still used today, with much better materials as well as devices, for improving ambulation. Today Hanger, Inc. is the largest deliverer of prosthetic devices.

Progress in prosthetic design dramatically accelerated following World War II, which produced a large number of young amputees, requiring the most advanced technology to restore their function. Engineers made a major and lasting impact on the field during that time, and today their involvement in prosthetic design is continuing to revolutionize the field. Perhaps the most significant progress was the result of collaboration between engineers and surgeons who designed procedures for optimal fitting and alignment of the prosthesis, which proved to be critical factors for optimal restoration of ambulation.

Following World War II, the large number of amputees spurred mass production of prosthetic devices. The U.S. Veterans' Administration began to systematically improve all aspects of prosthetic devices, and teamed up with Northrup Aviation, IBM, and others. New materials were developed for sockets, transitioning from wood and leather to plastic and metal. Many innovations in knees, feet, and hands were made [4]. The Thalidomide tragedy caused an even greater need for prostheses, especially for children.

In 1945, the *National Academy of Sciences* initiated a research program in prosthetics. Little was known about biomechanics at that time, so progress was slow. The early designs of the prosthetic socket were essentially cone-shaped wooden tubes, which supported the residuum, but offered little comfort and function, and no secure attachment. Improvements in structure, shape, and materials have resulted in greater comfort and fewer skin problems for amputees. Prostheses have since

1.4 Prosthetic Advances

advanced greatly; however they are still relatively poor substitutes for the natural limb.

1.4.2 Upper Limbs

About 500 years after the fourteenth century upper-limb (UL) prosthesis shown in Fig. 1.2, a more functional device, using body-powered cables for prehension, was patented, as shown in Fig. 1.5 [5]. This basic design is still in use today. One of the

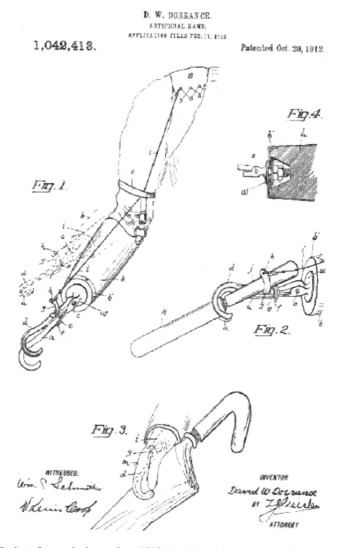

Fig. 1.5 Design of a prosthetic arm from 1912 that allowed for grasping. U.S. Patent of prosthetic arm by D.W. Dorrance, 1912 [5]

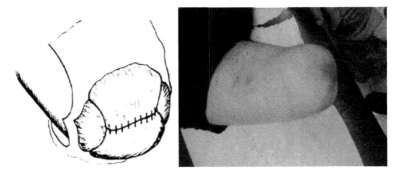

Fig. 1.6 Drawing and image of myoplasty, a surgical technique [7]

first powered hands was developed by IBM in the 1960s, and commercialized by *OttoBock,* a large orthopedic company. Today upper-limb (UL) prostheses can restore significant dexterity using advanced robotic technology. Progress for restoring mobility to lower-limb (LL) amputees has advanced from wooden peg legs to electromechanical legs with articulating knees and ankles. Exemplifying the most advanced LL and UL prostheses today are the *C-Leg* and *i-Limb*, respectively.

1.4.3 Surgical Techniques

Improved surgical techniques for amputation have played a big role in advancing limb restoration. Surgical procedures were greatly improved by the use of better anesthetics. In particular, the surgeons then had more surgical time to shape the residuum more appropriately for prosthetic use. When chloroform and ether became available as anaesthetic, more functional amputation residua were produced by designs to improve healing. An English surgeon, Dr. Edward Alanson, developed the technique of creating a conical-shaped residual residuum using skin flaps in 1782 [6].

Surgical techniques further improved the shaping and myoplasty of the residual limb in the 1950s, which allowed for better fit and attachment of the prosthetic socket (Fig. 1.6 Myoplasty). This type of surgery is depicted at left, and the healed residuum at right. Dr. Burgess also introduced the idea of applying a preparatory limb socket almost immediately after surgery to reduce swelling. This concept has led to the immediate postoperative prosthesis (IPOP), commonly done after amputation. The IPOP allows for quicker healing and provides a sense of security for the patient.

1.4.3.1 Myoplasty on the Stump

Prosthetic advancements are intimately tied to surgical advances. Surgical techniques and sites for amputation are influenced by the available prosthetic technology. Prior to the 1960s, most amputations were done at the trans-femoral level, i.e., above the knee (AK), even if the trauma or disease was primarily below the knee (BK); however, as surgical practice improved, outcomes improved as well, and BK amputation became preferable. A BK surgery can preserve knee function by re-attaching the knee flexor and extensor muscles onto the residual tibia. Another major surgical advance is the IPOP, which installs a preliminary prosthesis onto the residuum, which reduces swelling and gives the patient a feeling of security and hope for future functionality.

1.4.4 Causes of Limb Loss

Limb loss occurs through accidents, war, and disease. Traumatic amputations may result from automobile and machinery accidents, and freezing, firearms, electrical burns, and the misuse of power tools account for many as well. Amputations are also triggered by some accidents and conditions that cause limb paralysis, with the goal of replacing a useless limb with a functional prosthesis. Birth defects involving limb deficits are also responsible for amputation and artificial replacement. Finally, amputation is often performed in cases of disease, especially circulatory deficiency and infections, with diabetes causing most of the cases, and mostly involving the lower limbs.

As medical and surgical treatment for these traumas and diseases improves, the inevitability for amputation lessens. Often medical and surgical improvements have preserved many limbs that previously would have been lost. For example, infection, once a major cause of amputations, can usually be controlled by use of antibiotics. Moreover, severed limbs can sometimes be surgically re-attached.

1.4.4.1 Preserving Limbs and Limb Function

In cases of dysvascularity, usually related to diabetes, tissue damage accelerates when the foot loses sensation due to neuropathy. In the absence of sensation, walking produces abnormal foot pressures that damage tissue, causing ulcers that become necrotic and can threaten the entire body. While antibiotic drugs may control the infection and save the limb, or at least portions of it. If not cured, diabetic neuropathy often necessitates amputation of the entire limb.

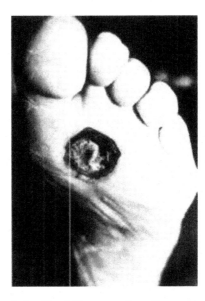

Fig. 1.7 Dysvascular foot lesion. Oandplibrary.org. https://www.google.com/url?sa=i&source=images&cd=&cad=rja&uact=8&ved=2ahUKEwizrPjpsbXmAhUGmeAKHVLNCYkQjB16BAgBEAM&url=http%3A%2F%2Fwww.oandplibrary.org%2Fpoi%2Fpdf%2F1993_03_189.pdf&psig=AOvVaw0uJ26syopTdmFj55v7Zoib&ust=1576421474907820 [6, 8]

If left unattended, a blister or small sore or even an infected ingrown toenail can quickly lead to tissue death and amputation. According to Michael Pinzur, MD, professor of orthopedic surgery at Loyola University Medical School in Chicago, foot ulcers are the most common reason a diabetic patient requires hospital admission. "At any point in time, 3–4% of the diabetic population will have a foot ulcer or a foot infection" (Fig. 1.7).

Other causes of LL amputation are tumors (e.g., osteosarcoma), trauma, congenital deformity, infection (e.g., osteomyelitis), and "dead, bad, or useless limb." Malignancy (cancer) may be a cause for amputation of either upper or lower limbs. With regard to the upper extremity, the most common cause of amputation is trauma, generally occurring in young males, either from war or accident. The level of amputation varies widely according to individual situations.

Congenital limb deficits, resulting in the absence of all or part of a limb at birth, are not uncommon. There was a spike in the incidence of limb deficits in persons born between 1958 and 1961 in the U.S., generally attributed to the in utero use of Thalidomide; however, other factors could be involved in these birth defects, since there is no certainty about the origin. The most frequent deficit of these defects is absence of most of the left forearm, which occurs slightly more often in girls than in boys; sometimes there is complete absence of parts or all of the four extremities. In such cases amputation may be indicated; however, decisions are difficult, since even a weak, malformed part is worth preserving if sensation is present and the partial member is capable of controlling some part of the prosthesis.

1.5 The Costs of Amputation

Amputation of limbs is very costly not only for the operation with costs >$60,000 in the 1990s, but also in terms of lost wages, productivity, and lifestyle adjustments. The most common amputation is of the leg, due to ulceration secondary to vascular failure associated with diabetes. While limb salvage can be attempted by surgical revascularization, it may at least serve to confine the amputation site to loss of a few toes rather than further up to a point where loss of the entire foot is necessary. According to Richard Chambers, MD, co-chief of the diabetes amputee program at Rancho Los Amigos National Rehabilitation Center in Downey, CA, "Even if it turns out that vascular reconstruction is not efficacious, there may be benefit nevertheless in the form of a patient who has a higher level of emotional acceptance". He or she may feel that at least everything was tried to save the foot. This helps the patient after the fact to avoid experiencing the lingering doubt that there still might have been some way to save the foot. "It's bad enough to have to lose part of your body, but it's far worse to be left thinking that it was an unnecessary loss. I believe that giving patients peace of mind is an important component of the treatment where amputation is concerned. At the very least, if a diabetic patient who consents to amputation can be made to appreciate the risks of surgery, it can avoid future surprises if the surgery doesn't go well or if the patient develops an infection."

1.6 Challenges for Lower-Limb Amputees

Locomotion is impossible without a leg, and even standing is uncomfortable without crutches. The loss of joints and the surrounding tissues disrupts the sense of feel and position, possibly triggering the troublesome "phantom sensations" of the lost limb. During ambulation, amputees must bear the total weight of the body on their residual limb, within the socket. The tight environment of the socket is not a hospitable place for muscle health, especially when they are no longer attached to the skeleton. As a result, residual muscles atrophy, and frequently develop sores. The limb-socket interface must be carefully designed to protect the residuum, while also transmitting body weight force efficiently.

Amputees face many challenges ambulating in the real world, even when using modern trans-tibial prostheses. There are many types of prosthetic designs for both the upper and lower limbs that serve users well, and improvements are continuing to be made with modern technology and engineering. Nevertheless, while current prostheses are capable of restoring a fairly high level of ambulatory competence, many amputees find coping with everyday activities challenging. The external environment is generally designed for sound-limb ambulators, for whom tasks like walking up and down stairs or on uneven ground, and in tight spaces, and running, dancing, swimming, and jumping are easy, natural, and seamless. Accomplishing these activities while wearing a leg prosthesis is challenging and substantially more difficult. Normal everyday activities require a high degree of functionality from prosthetic components, as well as expenditure of extramuscular energy from the residual limb and extra cognitive work for controlling the prosthesis. Amputees must learn adaptive strategies to navigate stair climbing, to overcome the limitations of prosthetic components, but these exert a cost in terms of physical effort and awkwardness. The forces experienced by the natural knee flexions during stair walking produce 1.31 Nm/kg of torque, whereas the artificial knee flexion moment for amputees is a small fraction of that (0.28 Nm/kg). This is mainly due to the absence of residual knee extensors and hamstrings [9]. Walking downstairs is even more challenging, since the knee tends to experience a large flexion moment on the down step, which can cause falls.

1.7 Types of Lower-Limb (LL) Amputations

Amputations are generally classified according to the level at which they are performed. Leg amputations range from below knee (BK) to above knee (AK), and most arm surgeries range from above elbow (AE) to below elbow (BE). More severe cases can involve shoulder disarticulation, and less severe can involve wrist disarticulation. The general acronyms used for lower-limb and upper-limb amputations are LL and UL. The general rule for surgical practice is to save all limb length that is medically possible. Amputation surgeries are designed to preserve as much limb length as medically possible.

1.7.1 Below-Knee (BK) Amputations

Any amputation below the knee joint is known as a below-knee (BK) amputation. Because circulation is often compromised in the amputated leg, the BK amputation is usually performed at the junction of the upper and middle third sections. Since the muscles that activate the shank remain intact, the surgery allows for nearly full use of the knee. Ideally, the BK residuum should not bear weight at the end of the stump, because the socket provides weight bearing throughout its length, as described in Chap. 4. Surgical procedures have tried to obtain weight bearing through the end of the below-knee residuum, but these have not been successful, due to tissue damage at the posterior surface. Trans-tibial (BK) is now the most common lower-limb [10] amputation due to perfection of surgical techniques, muscle stabilization, and the rigid postoperative dressing (Fig. 1.8).

Fig. 1.8 Artistic rendering of hip muscles following BK amputation

1.7.2 Knee Disarticulation

Complete removal of the lower leg, or shank, is a knee disarticulation, which allows for weight-bearing forces throughout the length, within the socket. Knee disarticulation is a difficult anatomy to fit into a prosthesis, and special surgical techniques are needed to overcome the problems posed by the length and shape of the residuum.

1.7.3 Above-Knee Amputations

Amputations above the knee (AK) and through the thigh require a socket that bears weight through the ischium, that part of the pelvis upon which a person normally sits. Body weight cannot be taken through the end of the residuum, since it is the soft and vulnerable tissue.

1.7.4 Hip Disarticulation and Hemipelvectomy

Hip disarticulation involves removal of the entire femur and requires a hip socket for support. With slight modification the same type of prosthesis can be used by the hemipelvectomy patient, that is, when half of the pelvis has been removed [11] (Fig. 1.9).

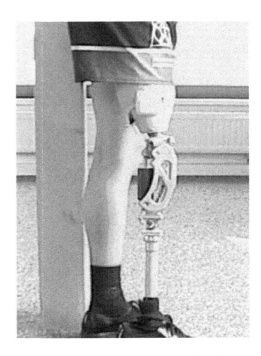

Fig. 1.9 Above-knee amputee with hydraulic knee. Carlson, The Lord Corporation

1.8 Upper-Extremity Amputations

1.8.1 Functional Comparison of Upper and Lower Limbs

Restoring the upper limb is more difficult than for legs because: (1) the motion repertoire of the leg is simpler and more predictable and (2) the lower limb is more essential to life than the arm. The upper limb, being relatively independent from the rest of the body, enjoys much freedom of movement, and almost infinite modes of movements. The lower limb, sandwiched between the ground and the torso, is obliged to the needs of the body as a whole; it cannot choose to support some parts of the body and not others or to walk with the body through only portions of each step. Motions of the knee and ankle are routine and predictable during each step, whereas upper-limb patterns are unpredictable.

1.8.2 Partial-Hand Amputations

When sensations of the hand are present, the surgeon can attempt to save functional parts of the hand in lieu of disarticulation at the wrist. Any method of obtaining some form of grasp, or prehension, is preferable to the best prosthesis. If the result is unsightly, the residuum can be covered with a plastic glove, lifelike in appearance, for those occasions when the wearer is willing to sacrifice function for appearance. Many prosthetists have developed special appliances for partial-hand amputations that permit more function than any of the artificial hands and hooks yet devised and, at the same time, permit the patient to make full use of the sensation remaining in the residuum. Such devices are usually individually designed and fitted.

1.8.3 Wrist Disarticulation

Restoring hand function after wrist disarticulation requires a special socket that reproduces the complex anatomy of the wrist joint. This technology enables the pronation-supination of the forearm—an important function of the upper extremity.

1.8.4 Below Elbow

Amputations through the forearm are commonly referred to as below-elbow (BE) amputations and are classified as long, short, and very short, depending upon the length of residuum. Residua longer than 55% of total forearm length are considered long, between 35 and 55% as short, and less than 35% as very short.

Long residua retain the rotation function in proportion to length; long and short residua without complications possess full range of elbow motion and full power about the elbow, but often very short residua are limited in both power and motion about the elbow. Range of pronation-supination decreases rapidly as length of residuum decreases; when 60% of the forearm is lost, no pronation-supination is possible. Devices and techniques are being developed to make full use of all functions remaining in the residuum.

1.8.5 Above Elbow

Amputations through the upper arm are referred to as above-elbow (AE) amputation. In practice, residua in which less than 30% of the *humerus* remains are treated as shoulder-disarticulation cases; those with more than 90% of the *humerus* remaining are fitted as elbow-disarticulation cases.

1.8.6 Shoulder Disarticulation and Forequarter Amputation

Removal of the entire arm is known as shoulder disarticulation but, whenever feasible, the surgeon will leave intact as much of the *humerus* as possible to provide stability between the residuum and the socket. When it becomes necessary to remove the clavicle and scapula, the operation is known as a forequarter, or interscapulo-thoracic, amputation. The very short above-elbow, the shoulder-disarticulation, and the forequarter cases are all provided with essentially the same type of prosthesis.

1.9 The Postsurgical Period

A technique for achieving optimal outcome, especially after LL amputation, is the immediate postoperative prosthetic (IPOP) procedure. The IPOP is a rigid plaster dressing placed over the residuum, which serves as a temporary socket, and the use of an adjustable prosthetic leg or arm which can be removed and reinstalled easily. The cast socket is left in place for 10–12 days, during which ambulation is encouraged. At the end of this time the cast socket is removed, the stitches are usually taken out, and a new cast socket is provided immediately. The original prosthetic unit is replaced and realigned. The second cast socket is left in place for 8–10 days at which time a new cast can be taken for the permanent, or definitive, prosthesis. The surgeon and others on his hospital staff will do everything possible to ensure the best results, but ideal results require the wholehearted cooperation of the patient.

The IPOP procedure:

- Reduces swelling and begin shaping the remaining limb.
- Increases chances for future prosthetic use and mobility.
- Improves outcomes and maximizes the value of rehab care.
- Protects the remaining limb from trauma.
- Preserves and improves range of motion and strength of the nearby joints and the entire body.
- Bridges between surgery and definitive prosthesis allows initial gait training.

As casting technology and materials have advanced, the practice of fitting a temporary limb soon after surgery, even before the patient has awoken from the anesthesia or after sutures are removed, has become standard practice. Ambulation training begins within 24 h in many cases. Results with many patients of all types have shown immediate postsurgical fitting of prostheses to be the method of choice when possible. Healing seems to be accelerated, postsurgical pain is greatly alleviated, contractures are prevented from developing, phantom pain seems to be virtually nonexistent, less psychological problems seem to ensue, and patients are returned to work or home at a much earlier date than seemed possible only a few years ago.

The IPOP has many benefits, including prevention of muscle contractures, which are difficult to correct. At first, exercises are administered by a therapist or nurse; later the patient is instructed concerning the type and amount of exercise that should be undertaken. The patient is also instructed in methods and amount of massage that should be given the residuum to aid in the reduction of the residuum size. Further, to aid shrinkage, cotton-elastic bandages are wrapped around the residuum and worn continuously until a prosthesis is fitted. The bandage is removed and reapplied at regular intervals—four times during the day, and at bedtime. The amputee is taught to apply the bandage unless it is physically impossible for him to do so, in which case some member of his family must be taught the proper method for use at home.

It is natural for the patient to feel depressed during the first few days after surgery, but after he becomes aware of the possibilities of recovery, the outlook becomes brighter, and he generally enters cooperatively into the rehabilitation phase.

1.10 Regulation of Prosthetic and Orthotic (P&O) Devices

No medical device can be commercialized in the U.S. without meeting the requirements of the Food and Drug Administration (FDA). Since most P&O devices are (up until recently) noninvasive, the level of regulation by the FDA is relatively low. Most P&O devices are, in fact, exempt from FDA regulation since they are applied noninvasively to disabled persons. Governmental regulations, however, change along with changes in technology, so the above regulations may become out of date.

1.11 Multidisciplinary Approach to Prosthetic Design

The field of prosthetics has seen more progress during the past few decades than in all the preceding 2000 years of limb-making. The rapidly emerging technologies and scientific breakthroughs are triggering ethical and socioeconomic implications for all medical practice, including limb restoration. New prosthetic options will not only include new hardware, but invasive interfaces, and bionic implants; these may not be affordable to many clients. The prosthetics market is relatively small and relatively poor in funding. Prosthetic devices may one day prove to be superior in function than natural limbs, giving some users with no medical need new bionic options [2]. Human-operated manipulandums are already available and being used to augment human performance.

The limbs move in response to systems of internal and external forces, and in accordance with the laws of mechanics. To restore to any satisfactory extent the functions lost through amputation of an extremity therefore requires intimate knowledge not only of the structure, form, and behavior of the normal limb but also of the techniques available for producing complex motions in substitute devices activated by residual sources of body power. Engineers are increasingly involved in prosthetic design and are becoming versed in sciences of biomechanics, surgery, anatomy, physiology, and rehabilitation. In the past 2 years, over 200 engineering research studies on prosthetics have been published.

1.12 Exercises

1. Look up the most advanced LL and UE prostheses you can find, and write a sentence on the capabilities of each. Cite the source, i.e., website or citation, and date.
2. State three advances in materials that have improved the functionality of limb prostheses.
3. How can knee function be preserved after BK amputation?
4. The first step in developing a model is to represent the problem as simply as possible, showing only the relevant parameters and variables. For example, analyze the reaction force at the residual limb for the case shown in Fig. 1.9. Please draw using simplified body segments and insert the correct arrows for forces and moments.

References

1. Ziegler-Graham K, et al. Estimating the prevalence of limb loss in the United States: 2005 to 2050. Arch Phys Med Rehabil. 2008;89(3):422–9.
2. Bennett Wilson A Jr. Artif Limbs. 1970;14:1–52.
3. Pare A, Oeuvres completes. Copy in the National Library of Medicine; 1840.
4. Saunders JBDM. Prelude, prophecy, and promise. O&P Library. Artif Limbs. 1955;2(1):1–3.
5. Dorrance D. Artificial hand. U.S. patent 1042418; 1912.
6. Tröhler U. Edward Alanson, 1782: responsibility in surgical innovation. J R Soc Med. 2008;101:607–8. https://doi.org/10.1258/jrsm.2008.08k011.
7. Mondry F. Myoplasty surgical technique. J Surg Orthop Adv. 1956;19(1):35–43.
8. Jernberger A. The neuropathic foot. Prosthet Orthot Int. 1993;17:189–95.
9. Schmalz T, Blumentritt S, Marx B. Biomechanical analysis of stair ambulation in lower limb amputees. Gait Posture. 2007;25(2):267–78.
10. Fairley M. Artistic rendering of hip muscles following BK amputation. Northglenn: O&P Library; 2013.
11. Carlson D. Personal communication; 2010.

Chapter 2
Human Limb Biomechanics

2.1 Introduction: Divisions of Biomechanics

Biomechanics applies Newtonian mechanics to living entities; it describes the effect of forces upon the form or the motion of bodies and has two subdivisions—statics and dynamics. *Statics* is the study of bodies at rest or in equilibrium as a result of the forces acting upon them. *Dynamics* is the study of moving bodies. *Kinematics* and *kinetics* are both concerned with moving bodies; *kinematics* deals only with describing their motion, and *kinetics* describes the forces causing the motion. To understand the workings of prosthetic devices, a knowledge of each of the following topics is required.

1. *Rigid Body Mechanics*

    (a) Statics
    (b) Dynamics

    - Kinematics: The study of *motion* without regard to the forces
    - Kinetics: The study of *forces* producing motion

2. *Deformable Body Mechanics*

    (a) Elasticity
    (b) Plasticity
    (c) Viscoelasticity

3. *Fluid Mechanics*

    (a) Viscosity, Reynolds number
    (b) Pneumatics, gases

2.2 Forces Within Living Limbs

All bodies on earth are subjected to external forces, mainly gravitational. All living bodies are also subject to three kinds of internal forces generated mostly from muscle: (1) *compressive* forces that tend to push or squeeze the body's materials together, (2) *tensile forces that* tend to pull materials apart, and (3) *shearing* forces that cause one part of the body to slide with respect to an adjacent part. These forces cause motions and deformations of the body, according to its structure and material properties. Consider the arm holding a weight as shown in Fig. 2.1 [1]. Each of the force types, compression, tension, and shear, are present; specific types of deformation or strain are also present. Compression tends to shorten the body and tension lengthens it, while both cause shear as a by-product. Axial loads to semi-rigid objects, including bone, muscle, and tendons, produce both stress and strain. As seen, the torque at the elbow is equivalent to the lever arm of 360 mm times the ball weight of 10 kg. For equilibrium, all forces and the torques must sum to zero. The muscle must exert a counter torque to the downward torques equal to F_b times the lever arm of 50 mm.

2.3 Stress on Limbs

A body, be it either animate or inanimate, deforms a specific amount according to its elastic modulus, when acted upon by an external force. The amount of deformation depends on the stiffness of the body, as measured by Young's modulus, and the magnitude of the force applied per unit area. Since forces are applied upon a specific area of the body, their magnitude is quantified as stress: the force magnitude divided

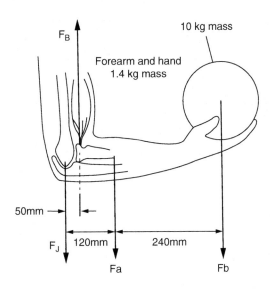

Fig. 2.1 Forces and torques on the arm. The force on the humeral joint is F_j; force of the biceps is F_a force of gravity on the arm center-of-mass; F_b is force of the ball. O&P Library > POI > 1979, Vol 3, Num 1 > pp. 4–12. J. Hughes, N. Jacobs

2.4 Free Body Diagrams

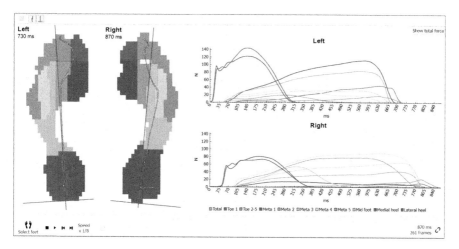

Fig. 2.2 Forces exerted on each foot while walking. Each color line on the plot corresponds to the anatomical area of shaded the same color of the foot on the map (left). https://www.google.com/url?sa=i&source=images&cd=&cad=rja&uact=8&ved=2ahUKEwiv1aeE2LXmAhUJUt8KHaDmAKkQjB16BAgBEAM&url=https%3A%2F%2Frsscan.com%2Fresources%2Fplantar-pressure-distribution-and-foot-kinematics-during-walking-and-running%2F&psig=AOvVaw26-cLBcGrl-ZQV7xt_LNLk3&ust=1576431580148336. (With permission from Rscan company)

by the area upon which it is applied. Stress may be static or variable in magnitude, time, and space depending on dynamics and the material properties. Stress is equivalent to surface pressure, an example of which is shown in Fig. 2.2. Here, foot pressures during walking are shown, with colors representing the magnitude of stress. As can be seen, the heel experiences the most force at the onset of striking the ground. Other aspects of the feet distribute the body force more evenly, producing less stress.

2.4 Free Body Diagrams

Replacing human limbs with artificial ones requires understanding of their motions and forces. A crucial design tool is the free body diagram (FBD) for designing the mechanical structures that will replace the missing limb. Human body components can be simplified (parameterized) where possible as struts, rods, and joints, with assigned physical and material properties. Parameters can be described in terms of kinematical dimensions of Length (L), Angle (Radians), and Time (T). To analyze the *forces* involved in motion, kinetics (or dynamics) is used. An example of rendering a physical situation with a FBD is shown in Fig. 2.1

While free body diagrams can vary in notation, the most useful diagram for a rigid body should follow these rules:

1. First, *sketch* the body in as simple form as possible, isolated from its surroundings. This can be in the form of boxes labeled for each limb or part. (You can add details later.)
2. *Isolate* the body from its external supports, and replace them with forces and force couples. You may "cut" the body wherever you need in order to see the forces at a particular location. You can leave in the external supports that may not be relevant to the problem (for example in Fig. 2.5 the ground support is still shown, without a force). The forces you should show are: (1) externally applied, (2) reactions at supports or points of contact with other bodies, (3) gravity forces affecting the components. Any forces that are internal to the body should not be inserted, since these forces are always arranged in an equal and opposite couple.
3. Insert and label the distances, mass centers, and other dimensions that can be relevant to solving the problem.

The FBD in Fig. 2.1 can be depicted in both anatomical and stick figure (just bones) forms as shown in Fig. 2.3. More relevant analysis is the case of forces applied to the human arm holding a ball, as depicted below.

The three downward (clockwise) force vectors acting on the arm represent the weight of the arm at the center of mass (G), the ball (W), and the reaction force at the elbow. The upward force B represents the Biceps. The downward forces, G and W, multiplied by their respective lever arms, L_g and L_w, shown as arrows from the elbow joint tend to rotate the arm in a clockwise direction, while the upward force, B, multiplied by its lever arm, L_B, tends to counteract the downward torques. Thus, each force exerts a torque or "moment" equal to their magnitudes times their lever arms, and counteracted by the torque induced by the Biceps muscle times its lever arm, $L*B$. If $L_g * G + + W * L_w = L_B * B$, the arm is static; otherwise the arm will rotate.

As articulated by Newton: In order for a body to remain at rest (fixed, relative to a point in space) the vector sum of all forces acting upon it must be zero. **Force and moment Equilibrium is a** static situation and exists when all the forces and moments applied to an object sum to zero, i.e., there is both "**force equilibrium**" and "**moment equilibrium**."

2.4.1 Balance of Forces and Torques

A body is in equilibrium if all forces and torques applied to it add up to zero, as depicted in Fig. 2.3. The top panel depicts an arm holding a ball, and the corresponding free body diagram is shown at bottom. **A more anatomically detailed depiction of the arm holding a weight is shown in** Fig. 2.3.

2.5 Kinematics and Kinetics

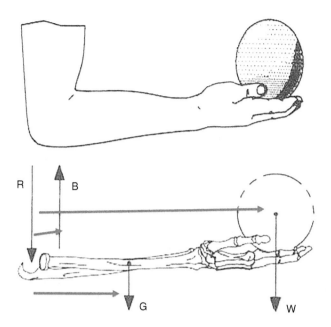

Fig. 2.3 Anatomical renderings of arm holding a ball. (**a**) Top depicts an intact arm holding a ball of weight, "*W*." This is a static situation. Lower figure shows a free body diagram of forces exerted on an arm holding a ball. "*G*" represents the weight of the arm at the center of mass, "*R*" represents the reaction force at the elbow, and "*B*" represents the force of the biceps. Below is the free body diagram showing the forces involved in the sketch. Note that for equilibrium the upward forces and torques need to balance the downward forces, so the net forces add up to zero. https://www.google.com/url?sa=i&source=images&cd=&cad=rja&uact=8&ved=2ahUKEwjwu6X50rXmAhVnleAKHULBBMgQjB16BAgBEAM&url=https%3A%2F%2Fwww.ecronicon.com%2Fecor%2Forthopaedics-ECOR-01-00001.php&psig=AOvVaw0LQrmTzvGSNBre_rl-GMap&ust=1576430263562412 *Copyright:* © 2015 Mohammad Kamran Shahid. *et al.* This is an open-access article distributed under the terms of the Creative Commons Attribution License, which permits unrestricted use, distribution, and reproduction in any medium, provided the original author and source are credited

2.5 Kinematics and Kinetics

The most common major motions of the human body are depicted in Fig. 2.4. This standard anatomical terminology can precisely describe most human motions. The fingers of the hand can also produce most of the motions described here.

Most bodily motions usually combine two or more of the basic movements shown, producing a large movement repertoire of regular, predictable patterns listed below.

1. Vibration or oscillation of a segment, bone, or structure
2. Back-and-forth motions, such as angular displacement of joints or linear displacement of muscle fibers
3. Articulated motions of the fingers or toes

https://www.researchgate.net/publication/26710808_Redirection_of_center-of-mass_velocity_during_the_step-to-step_transition_of_human_walking

Kinematics, as depicted in Fig. 2.4, relates only to the motion, without regard for the forces involved. Kinetics describes the forces involved in the motion. The top left panel in Fig. 2.5 shows a trajectory-type motion starting from point zero and ending at point 1 m, after 5 s. The motion describes a smooth curve showing relatively rapid start and stop, with a moderate speed in between. This kinematic behavior is typical of the movement of human limbs. As can be seen in the middle left figure, velocity peaks midway in the movement, and in the bottom left figure, acceleration reverses at mid-movement. The kinetics, sometimes referred to as dynamics, show the velocity and energy versus time.

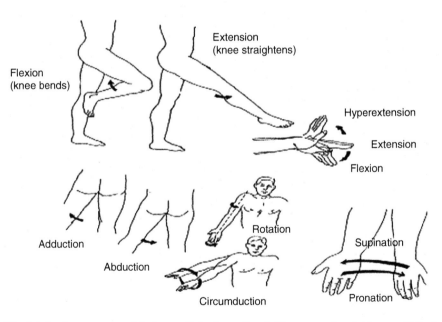

Fig. 2.4 Basic movements of the human limbs. https://brooksidepress.org/brooksidepress- "All of our products are freely available online for downloading and use in the furthering of medical education"

2.6 Dynamic Analysis

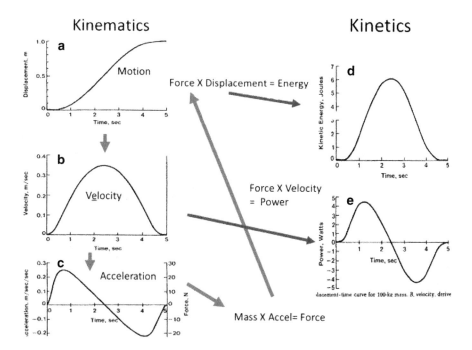

Fig. 2.5 Kinematics and dynamics of limb motion

The kinematical and kinetic data in Fig. 2.5 typify the motion of the hand during a reach. In Fig. 2.5a, the linear displacement of the hand as a function of time is shown. The movement took about 5 s to complete, and the curve approximates a sigmoid. In Fig. 2.5b (see arrow), the velocity of movement is shown. Note that it has a bell shape, with slowing velocities at the beginning and end, and maximum velocity at the middle of the movement. In Fig. 2.5c, the acceleration trace shows a quasi-sinusoid, with acceleration crossing zero at the peak velocity. If the mass of the moving structure is known, the forces, energy, and power involved can be calculated according to Newton's second law, as depicted in Fig. 2.5d, e.

2.6 Dynamic Analysis

Dynamics is the analysis of forces producing motions.

The net torque on a joint is due to the resultant force from all muscles acting on the joint, times their lever arm. The resultant force is the vector sum from all the muscles, as depicted in Fig. 2.6.

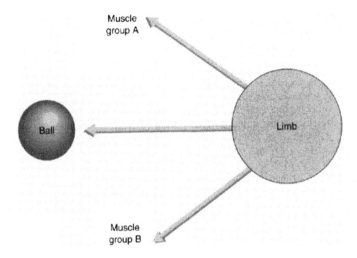

Fig. 2.6 Net force on a joint

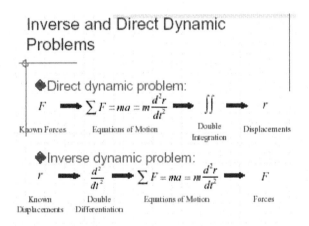

Fig. 2.7 Inverse and direct dynamic problems

Forward, or direct, dynamics is the process of computing the motion trajectory (kinematics) when the forces are known. *Inverse* dynamics is a process that computes the *forces* acting on the body when the motion trajectory is known. Both use Newton's laws, as depicted in Fig. 2.7.

Every directional movement humans make requires a neuromotor system to solve the direct dynamic problem. As depicted, to reach a target, "r," our muscles apply forces, "F," to the appropriate joints that place our "end effector" (i.e., hand) at the desired location, "r." The forces from diverse muscles must be coordinated and adjusted during the motion via sensory feedback to achieve the correct accelerations and decelerations to arrive at the desired endpoint.

2.7 Uniaxial Stress and Strain

The inverse dynamic problem is characteristic of robotic motions: to reach a particular displacement, r, the robot accelerations must be calculated in advance, and from these, the equations of motion are computed to determine the forces required to reach endpoint, "r." The reason why robotic motions must solve the dynamic problem inversely is because robots lack the continual sensory feedback which characterized human motion. Human movements can be continuously and finely controlled during a motion by feedback from the neuro-sensory motor system. Robots, in contrast, do not possess (yet) the fitness and adaptability enjoyed by humans.

2.7 Uniaxial Stress and Strain

Perhaps the most fundamental property of any structural material is its strength. To measure it, a representative specimen of the material, with a simple shape such as a bar or rod, is pulled until it breaks. Thus applying constant forces, F, to both sides of the bar as shown in Fig. 2.8 would translate into the stress, $\sigma = F/A$, assuming the forces were distributed evenly over the cross-sectional area, A.

When a bar of length L_0 undergoes linear stress, σ_x, in the x direction, it lengthens in that direction while shortening in the orthogonal (y) direction. The strain, ε_x, is:

$$\varepsilon_x = \frac{L - L_0}{L_0}$$

The resulting stress-strain curve is shown in Fig. 2.8. It is seen that for low strains, the slope (rise/run) is linear, until the yield strength is reached, at which point the rate of stress increase begins to slow, until the material reaches its ultimate

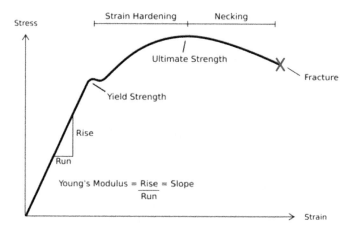

Fig. 2.8 Stress-strain curve. https://en.wikipedia.org/wiki/Stress–strain_curve

Mechanical Loading of Bone

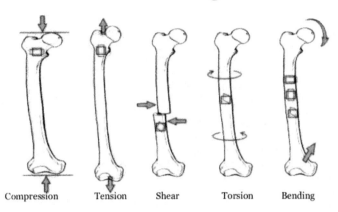

Fig. 2.9 Mechanical loading of bone. Modes of force loads on bone—Slideshare

tensile strength. After this point, the material experiences negative stiffness and fractures. These data provide two important parameters: material stiffness, as defined by the initial slope (rise/run), and the ultimate tensile strength (UTS). Material stiffness is thus defined as Young's modulus: $Y = \frac{\sigma_x}{\varepsilon_x}$.

The elastic modulus applies to both tensile and compressive loads. Testing is done on specimens of standardized size and shape. Stress in the specimen is calculated as the force divided by its cross-sectional area. The *energy* the specimen absorbs to failure is calculated as the area under the curve (AUC). Under these conditions the cross-sectional area of the specimen is known, and it is assumed that the force is uniformly distributed across it. After applying the force, the ultimate tensile or compressive strength of the material can be easily calculated from the stress-strain curve as shown in Fig. 2.9.

2.8 Beam Analysis of Structures

The bony limbs behave similarly to beams, in terms of strength and stiffness, and beam analysis can quantify their behavior. As seen in Fig. 2.9, bones are subject to a variety of loading types, including, compression, tension, torsion, shear, and bending. For example, axial force on a rectangular beam causes compression, bending, and tension *on the convex side and compression on the concave side. This tension is given by the formula*

$$\sigma = My / I$$

where sigma is stress, *M* is the force, *y* is the lever arm, or curvature in the beam, and *I* is the moment of inertia of the beam.

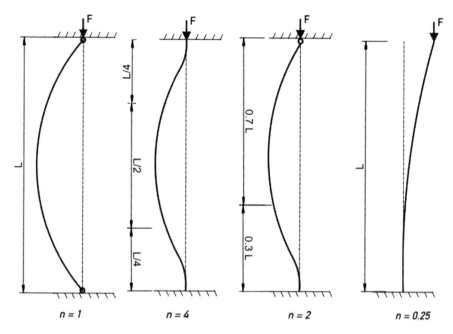

Fig. 2.10 Straight rods subjected to axial loads. Depending on the attachment type, the force, P, needed to buckle the beam is calculated using one of the equations above. (Engineering Toolbox 2012)

2.9 Beam Buckling

Sometimes, beams buckle, as shown in Fig. 2.10. The force, P, which makes them buckle is dependent on the type of attachments the beam has. A beam can be unattached at both ends, which means it is free to slide on its supports, like the one on the left, or it can be attached at both ends, like the middle one, or attached at one end, like the one on the right. As shown below, the buckling forces, "P" vary according to the type of attachment.

$$P_1 = \frac{\pi^2 EI}{L^2} \quad P_2 = 4\frac{\pi^2 EI}{L^2} \quad P_3 = 2\frac{\pi^2 EI}{L^2} \quad P_4 = \frac{\pi^2 EI}{4L^2}$$

2.10 Applied Stress to Cube

Mechanical testing of a cube-like structure can be done either uniaxially or biaxially. A uniaxial stress on a cube can be applied if the cube is constrained in one dimension. A biaxial strain will occur under a stress on an unconstrained cube. The simplest depiction of stress is in an unconstrained isotropic cube, as shown in

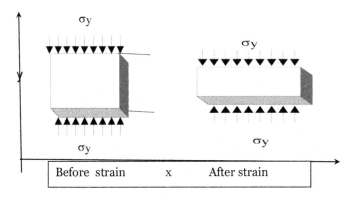

Fig. 2.11 Stress applied to an unconstrained cube

Fig. 2.11. Here, there is strain in both the *x* and *y* axes. When the cube is constrained, the stress-strain relationship is altered.

The strain in the *y* direction is: $\varepsilon_y = \dfrac{\sigma_y}{E}$

Because the transverse direction is unconstrained:

$$\sigma_x = 0 \quad \text{and} \quad \varepsilon_x = \nu \varepsilon_y$$

Now, consider the case where the *x* direction is constrained from movement. That is, transverse movement is resisted, making:

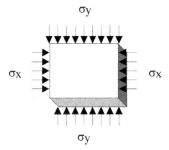

To prevent strain in the *x* direction, a stress σ_x must be applied:

$$\sigma_x = \varepsilon_x E = \nu \varepsilon_y E$$

In other words, to prevent transverse strain, you must push on the object with σ_x. This causes an added stress to the *y* direction, due to a tendency to strain:

$$\varepsilon'y = \nu \varepsilon_x = \nu^2 \varepsilon_y$$

2.11 Exercises

and

$$\sigma'y = Ev^2\varepsilon_y = v\sigma_x$$

Thus the new stress in the y direction is the original unconstrained stress plus the stress caused by transverse constraint:

Solving for σ_y we have the biaxial stress equation: $\sigma_y = E\varepsilon_y + v\sigma_x$

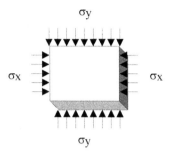

Terminology

Stress	The amount of force applied over a given area.
Strain	The amount of deformation for a given length.
Young's modulus	Stress/strain: the amount of force to obtain a certain amount of deformation. Synonymous with stiffness.
Ultimate tensile *Strength (UTS)*	Maximum force applied before a fiber breaks.
Yield strength	Maximum force applied before permanent deformation occurs.
Bending stiffness	Modulus during bending.
Brittleness/ ductility	Relative terms describing how much plastic deformation occurs before fracture; brittle = small deformation, ductile = high deformation.

2.11 Exercises

1. Draw an FBD for Fig. 2.1, assuming force equilibrium. Is this a static or dynamic situation?
2. Write the equilibrium equation for the following FBD:

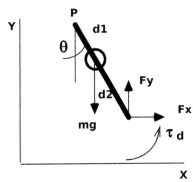

3. Using information in the appendix, compare the buckling force for a solid and a hollow pylon, both Titanium, with the following dimensions: Length = 25 cm, O.D. = 2.5 cm, I.D. (of hollow pylon) = 2 cm. Compare their strength-to-weight ratios.
4. Which type of beam supports is most likely to buckle the quickest (with least force)? Refer to Fig. 2.10

Reference

1. Hughes J, Jacob N. Normal human locomotion. Prosthet Orthot Int. 1979;3(1):4–12.

Chapter 3
Metrics and Mechanics of Human Limbs

3.1 Introduction

Restoration of limb function requires a custom-made prosthesis that attaches comfortably and securely to the residuum. This attachment between the residual limb and the prosthesis is known as the socket, which is discussed further in Chap. 6. Successful mating of residuum with prosthesis requires anthropometric data on the client: measurements of the client's residuum and relevant bodily features, and physiological characteristics that might influence operation of the prosthesis. The measurements are needed in both static and dynamic conditions, and, in the case of lower-limb restoration, the ambulatory capability and goals of the user. Relevant metrics, at a minimum, include dimensions and tissue properties such as stiffness, strength, circulation, and residual neuromuscular function. Metrics need to be assessed in both static and dynamic conditions in order to simulate daily conditions. Many tools are available to register these metrics, ranging from tape measures to digital imaging and computational models. Analyzing and interpreting kinematic and dynamic data is an important design tool that is discussed in this chapter.

3.2 Anthromorphometry

The human body has no straight lines: all its components are irregular and variable both within and among individuals, and over time, at both short and long scales. Equally variable are bodily material properties, including tissue density, stiffness, thickness, and health. These properties must be known as accurately as possible prior to installing the prosthesis, using manual and technological means.

Bodily dimensions are recorded in relation to identifiable landmarks, which are relatively stable points on the adult skeleton, depicted in Figs. 3.1 and 3.2. Bony sites are more stable than soft tissue and can be readily discerned by palpation.

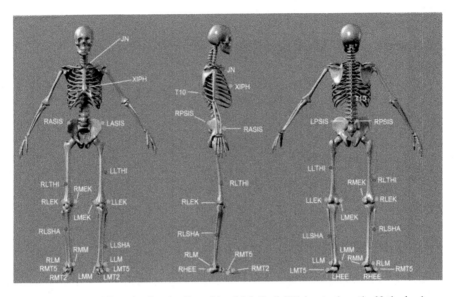

Fig. 3.1 Anatomical bony landmarks. From Motek Medical, BV Amsterdam, the Netherlands

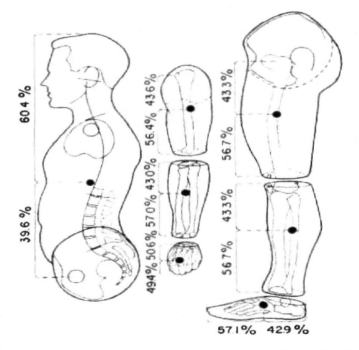

Fig. 3.2 Major body segments, showing the relative locations of vertical CoMs. Table 1 (Appendix) shows absolute data. Harless (O&P Library, no copyright). The relative positions of the CoMs of the human body. Average masses of segments are listed in Table 2 of the Appendix. From the O&P Library

Physiological and material properties of limbs and relevant anatomy are similarly recorded in relationship to baseline normative data, where available. Relative dimensions and masses of body segments are shown in Fig. 3.1 and Table 1 of the Appendix.

Anthropomorphic data on human limbs are available from a few studies, tabulated in the Appendix. Although the idea of a "normal" body is a fallacy, some of the measurements have established the "standard man" depictions, such as seen in Figs. 3.3 and 3.4. It should be noted that these data may not be representative of the current population with its wide range of age and difference of body build.

For biomechanical analyses, major limb structures can be simplified by splitting them into fixed segments as shown in Figs. 3.2, 3.3, and 3.4. In Fig. 3.2 the body is shown split up into seven segments: (1) head and torso, (2) upper arm, (3) forearm, (4) hand, (5) thigh, (6) calf, and (7) foot. Each segment is described by a center of mass (CoM) and weight distribution in percentage. The CoM is located by its relative vertical position in the sketch. This depiction assumes that the body is symmetrical along the midline, and left and right limbs are identical. This assumption is almost never correct for any individual. These data were obtained from measurements of absolute segment lengths of a group of males and then normalized to the entire segment length.

The anatomy of the human body can be represented in various ways, each of which is useful for illustrating its functions Fig. 3.2. A stylized rendering of the main body segments as rectangles is shown in Fig. 3.3. There is a frontal standing view and a walking view in the sagittal plane.

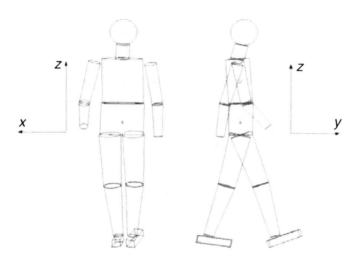

Fig. 3.3 Standard body showing principal segments. *International Society for Biomechanics*

Fig. 3.4 Human Body Axes. https://doi.org/10.1016/0021-9290(95)00017-C. **License Number** 4862020806643

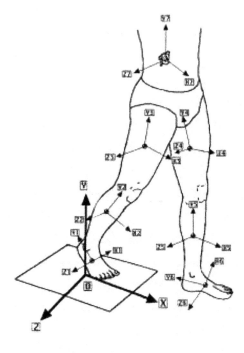

Fig. 3.5 Simplified walking coordinates

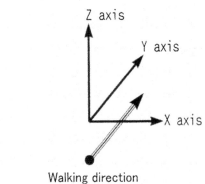

A three-dimensional coordinate standard of the human body established by the *International Society for Biomechanics* is shown in Fig. 3.4. Here the coordinates of individual segments, X, Y, Z, are shown at their respective center of mass. This stylized depiction shows a person standing (frontal view) and ambulating (sagittal view). Note the lower limb is composed of three segments and three joints. The entire body, including the hands (not shown), comprises 16 segments.

A simplified coordinate system that is commonly used for gait kinematics is shown in Fig. 3.5.

A simplified version of the walking leg is the pendulum, which applies during the swing phase as shown in Fig. 3.6. The simplifying assumptions are leg is a rigid

Fig. 3.6 Schematic view of pendulum model. *Annals of BME (Springer Scott Delp).* Simplified depiction of a swinging leg, with center of mass represented by the massive bob

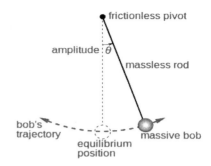

Fig. 3.7 Measuring arm volume. The volume occupied by water between any two levels is found by taking the difference between heights of water levels and applying suitable area factors. To find the volume of the forearm the displacement volume is found for the wrist to elbow levels in the lower tank and between the corresponding levels in the upper tank. The difference between these two volumes is the forearm volume. From the O&P Library (no copyright)

beam, with mass concentrated at the center of mass, and swinging on a frictionless pivot.

3.3 Measuring Body Segments

Segment volumes can be crudely estimated by measuring the volume of water displaced by the segment in a water tank, as shown in Fig. 3.7. A procedure developed at NYU has the subject place his segment in an empty tank which is subsequently filled with water. In this way, the subject is comfortable during the test, and the segment remains stationary to ensure stable results. A variety of tanks for the various segments—hand, arm, foot, and leg—are used. There are more modern methods available today using image scanning instruments.

Fig. 3.8 Manually measuring circumferences and tibial length of a leg residuum. O&P Library

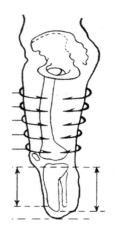

Measurement of the residuum is crucial to prosthetic performance. One of the major goals of the socket is to provide an intimate fit with the limb, so that the socket becomes a "total contact" fit. The exact shape as well as biomechanical features such as regional tissue stiffness must be measured.

Measuring the contours and volume of irregularly shaped body parts, such as a residual limb, can be done with a combination of tape measurements, photogrammetry, or digital imaging, which are discussed in Chap. 10. For example, the contours and landmarks of the BK residuum (Fig. 3.8) are measured with tape measure, as well as modern techniques such as laser scanning. Modern 3D printing has made the shaping of sockets a more scientific process.

3.4 Anatomical Coordinates

Fundamental to describing kinematics is a consistent 3-axis coordinate system. The axis is defined as an invisible line around which all motion takes place. Positioning of the axis will dictate how much motion will occur. The standard coordinates of x, y, and z, as shown, lie in the planes, horizontal, frontal, and sagittal, planes, respectively, as depicted in Fig. 3.9.

3.5 Measuring Human Movement and Dynamics

Understanding of the kinematics and kinetics of human limb control is simplified by the fact that many activities of daily life (ADL), such as walking, eating, self-care, and exercising, are relatively predictable, and hence restorable by fairly simple

3.5 Measuring Human Movement and Dynamics

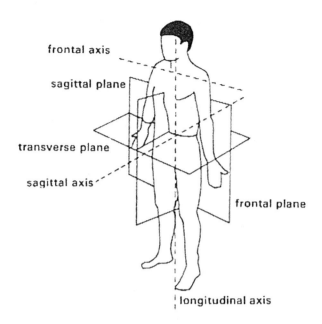

Fig. 3.9 Axes and planes of the human body. https://images.search.yahoo.com/search/images?p=human+body+axes+and+planes+diagram&fr=mcafee&imgurl=http%3A%2F%2Fi19.servimg.com%2Fu%2Ff19%2F14%2F59%2F63%2F40%2Fbodyax10.jpg#id=5&iurl=http%3A%2F%2Fi19.servimg.com%2Fu%2Ff19%2F14%2F59%2F63%2F40%2Fbodyax10.jpg&action=click

mechanisms. The most relevant kinematic variables of the human body are the angular ranges of its joints: the neck, shoulder, elbow, wrist, hip, knee, and ankle. While joint motions are largely rotational, with a small, variable amount of linear motion as the joint surfaces slide, their goals are all the same: positioning the end effector at a desired location. In gait, the goal is placing the feet correctly, and manipulation requires positioning the hand at the desired position. For each joint, it is important to know their paths of motion, linear and angular displacements, amplitudes of motion, and their velocities and accelerations. These are kinematic variables, which are concerned only with the motions, not the forces that caused them.

One of the first kinematical studies of gait was done by Muybridge who filmed a galloping horse. Prior to the era of moving pictures, there were questions whether or not horses actually "flew" during a phase of gait, meaning that all four hooves left the ground at some phase in the cycle. Muybridge tested and proved this phenomenon using a crude motion picture camera. This experiment showed gait to be a regular, reproducible pattern, and ushered in the use of motion pictures in the study of biomechanics (Fig. 3.10).

Fig. 3.10 Muybridge strobes. Twenty-four cameras spaced around the track were each triggered by the breaking of a tripwire on the course. Results showed that a horse's feet are not, as hitherto believed, outstretched, as if like a rocking horse, but bunched together under the belly. The horse is airborne in the third and fourth frames. https://en.wikipedia.org/wiki/The_Horse_in_Motion#/media/File:The_Horse_in_Motion_high_res.jpg

3.6 Human Walking Kinetics

Today, kinematic analyses can be obtained from a variety of sensors, such as inertial measurement units (IMUs), which are commonly present in cell phones.

Normal walking involves the coordinated, repetitive motions of all 16 body segments, shown in Fig. 3.3. To understand the physics of walking, the angular momenta of each segment are measured in walking subjects. While subjects walk across a force plate, as depicted in Fig. 3.11, IMUs on each segment record the inertias of each of those segments. From these data, combined with anthropomorphic data on the mass distribution of each segment: limbs, trunk, and head, the angular momentum of each limb segment can be calculated.

Results of walking analysis studies have shown that each swinging body segment produces a large angular momentum, but the summed angular momentum of the whole body (all the segments) approximates zero throughout a walking cycle at steady speeds. This somewhat paradoxical result can be explained by the motions of the segments, which are bilaterally reciprocal, so that the opposing limb movements in three dimensions (see Fig. 3.4) tend to cancel out each other's momentum. Thus, the overall whole-body angular momentum is minimized, which has important implications about energy usage, since opposing limb movements ideally reduce the

3.6 Human Walking Kinetics

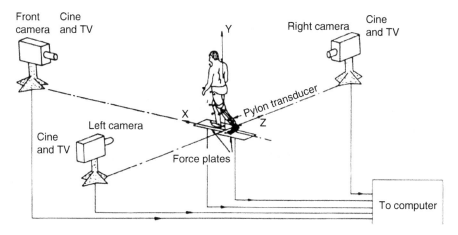

Fig. 3.11 Recording prosthetic walking kinematics and dynamics

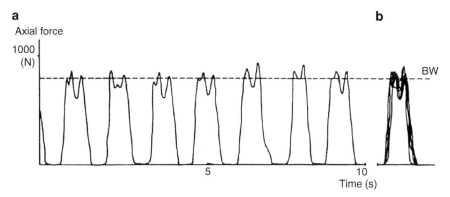

Fig. 3.12 The axial force measured by the pylon transducer during seven steps is shown. Each trace represents a single prosthetic foot strike. As seen in (**a**), each step produces a slightly different pattern, with two major peaks, one for heel strike and the following one for toe-off. Superimposition of all steps is shown in (**b**). Note the variability in force patterns among the steps. M. S. Zahedi *, W. D. Spence *, S. E. Solomonidis *, J. P. Paul *, O&P Library > POI > 1987, Vol 11, Num 2 > **pp. 55–64**

metabolic cost of walking. These data are important information for the optimal design of prostheses, in terms of the masses and moments of inertia of each segment.

The axial forces produced during each step by a person walking on a prosthetic leg are shown in Figs. 3.11 and 3.12. Note the peak axial forces are in the 900 N range and vary from step to step.

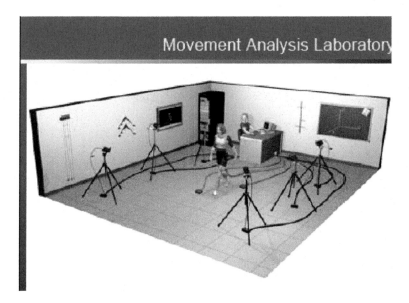

Fig. 3.13 *Modern methods of tracking human motion*. Modern technology, as seen in this figure, uses body-worn sensors, floor sensors, and cameras. A typical gait analysis laboratory employs 8–12 cameras, with force sensor plates in the floor to record forces in three dimensions. Stephen J. Stanhope, National Institute of Child Health and Human Development, NIH

3.6.1 Modern Methods of Tracking Human Motion

There are many methods for tracking human motion, including body-worn sensors and cameras. A simple method used by prosthetists is visual tracking of clients wearing their prosthesis without special markers. These trained observers can learn much about the special designs that are optimal for each client. With technology, computational methods can also be used to design optimal systems that are sometimes superior than human visual observations. For detailed tracing of all joints, motion analysis labs are used as shown in Fig. 3.13.

3.6.2 Measuring Joint Dynamics

Limb Joints are rotated by the lever action of muscles about the joint, which produces "moments." These are torques which tend to change the joint angle, either by flexing or extending the limb. For example, "ankle moment" refers to the muscular action about the ankle joint which may cause either plantar flexion or dorsiflexion. The moments are net external moments, normalized to % body weight times height. The physical meaning of these moments depends on the type of joint on which they are acting. For example, the net plantar flexion moment at the ankle is the moment sustained by the *dorsiflexor* muscles at early stance to control foot motion between heel strike and foot flat, providing a smooth lowering of the foot to the floor. The net

3.6 Human Walking Kinetics

Fig. 3.14 Joint moments on a walking leg.
J. Hughes *N. Jacobs *
O&P Library > POI >
1979, Vol 3, Num 1 >
pp. 4–12

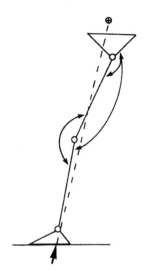

dorsiflexion moment is produced by the plantar flexor muscles of the foot at push off. At the knee, an extension moment is produced by the *quadriceps* to stabilize the leading limb at heel strike and prevent knee collapse. As the leg progresses over the foot during mid-stance, the *hamstrings* produce an external flexion moment. These internally produced moments, combine with external moments to produce the net forces acting on the articular surfaces.

A simplified sketch of an entire leg, with hip, knee, and ankle joints, during walking illustrates the opposing moments occurring at the joints, as shown in Figs. 3.14 and 3.15. Here we see that the upward force (the ground reaction vector) will cause an extension moment at the knee and a flexion moment at the hip. There is also a small plantarflexion moment at the ankle (Fig. 3.15). Depending on the exact direction of the GRF, these two moments will tend to cancel out the total moment of the leg. All joints in the body experience opposing moments during activities, so that the total momentum in a moving body can be close to zero, as described in Sect. 3.6.

3.6.2.1 Ankle Dynamics

The push-off phase of gait is assisted mostly by the ankle plantar-flexion, as shown in Fig. 3.15. As the calf muscle contracts, it produces an upward force, F_m, to pull the heel up and push against the ground reaction force vector, F_g. There is an effective mechanical advantage (EMA) acting about the ankle joint, equivalent to $t*R$, which works against the gravity torque, F_g*R.

> JSTOR is part of ITHAKA, a not-for-profit organization helping the academic community use digital technologies to preserve the scholarly record and to advance research and teaching in sustainable ways.

3.6.2.2 Interpreting Prosthetic Gait Data: Ankle Joint Moments

Restoring gait for a lower-limb amputee requires replacing the function of the lower-limb joints. This task is particularly complicated for the foot-ankle complex. The ankle must tolerate large forces and torques in all three axes during footfall, and its artificial replacement must be tailored according to the individual user. There are many different varieties of prosthetic ankles available that can be tested. To measure and optimize the ankle, their gait motion is analyzed in setups as shown in Fig. 3.13. In order to align the ankle optimally with respect to the leg, gait trials are done. The typical test bed includes sensors of joint angles (goniometers) and a force plate to register the magnitude of the forces during gait, as shown in Figs. 3.13 and 3.16. Typical data obtained on a subject with above knee amputation, wearing a prosthetic ankle, is shown in Fig. 3.16a. The subject was instrumented with IMUs and force sensors, which recorded the three sets of separate traces shown. These recordings are used to find an optimal alignment for the ankle prostheses. Thus the force results from alignments 1, 2, and 3 are shown on the same graphs.

The moment (torque) magnitudes experienced by the prosthetic ankle during walking in several planes with sensors are shown in Fig. 3.15. The recorded traces of medio-lateral ankle moments (Fig. 3.16c) during steps appear as thick lines because they are superimposed records from multiple individual steps with the leg prosthesis. Thickness of the lines relates to the variability among the steps. The top traces represent the ankle moment in the medio-lateral direction. Note that the largest moment occurred for alignment 2. Each of the traces shows a brief negative

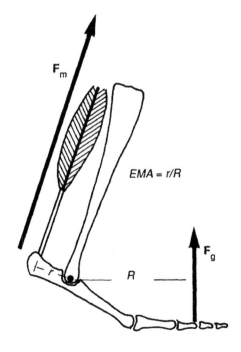

Fig. 3.15 Schematic illustration of the extensor muscles acting about the ankle joint to push off the ground. http://www.jstor.org. Tue Aug 28 12:05:42 2007

3.6 Human Walking Kinetics

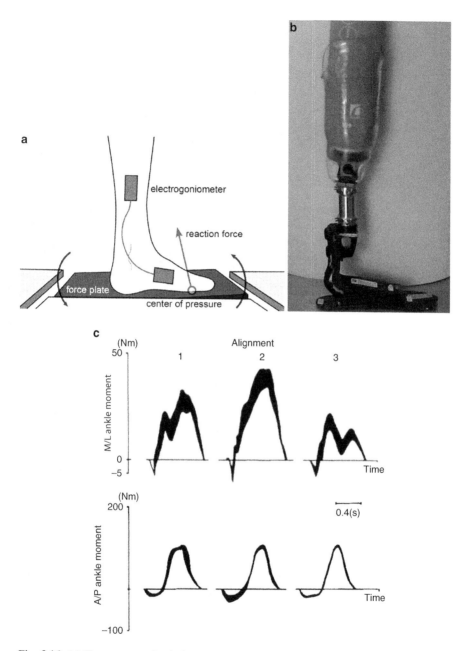

Fig. 3.16 (a) Top: sensors of gait force during gait. Robert D. Gregg, Elliott J. Rouse, Levi J. Hargrove, Jonathon W. Sensinger PLoS ONE 9(2): e89163. doi: 10.1371/journal.pone.0089163. (b) Bottom: Test Prosthesis. Arezoo Eshraghi, Noor Azuan Abu Osman, Mohammad Karimi, Hossein Gholizadeh, Ehsan Soodmand, Wan Abu Bakar Wan Abas. PLoS ONE 9(5): e96988. doi: 10.1371/journal.pone.0096988. (c) Effect of alignment on prosthetic ankle moments in two planes during gait

moment, as expected, at heel strike. The lower row of traces shows the anterior-posterior moments that are depicted anatomically in Fig. 3.14. One noticeable feature is the variation in the traces, as indicated by their thickness. It is apparent that the anterior-posterior traces are more repeatable, showing less variation, than the medio-lateral traces.

3.6.3 Muscle Activities During Gait

3.6.3.1 Monitoring Muscle Activity with Electromyography

Gait kinetics can be measured by muscle amplitudes and timing. Combined with anthropomorphic data and ground reaction forces, joint moments can be computed by applying Newton's laws to the centers of mass (CoMs) and the mass moments of inertia (MoIs) of the body segments. From these data, linear muscle forces can be estimated.

Electromyography (EMG) is employed to observe the sequencing, magnitudes, and coordination of leg muscles during gait. Such analysis reveals the relative roles of specific muscles in moving body segments. A sample EMG analysis of gait is shown in Fig. 3.17. Surface EMGs (sEMGs) are recorded by electrodes placed on the skin, near the center of relevant leg muscles. The raw EMG records are processed by amplification, filtering, and thresholding. Signal processing removes the high frequency components of the signal, as well as noise. For ease of interpretation, thresholding removes the background muscle activity, showing only the larger signals. Noise and lower amplitude signals are thus removed to highlight the main muscle activities. The processed EMG signals now appear as black, irregular blotches that demarcate periods of significant muscular activity. Note the walking man above begins his right heel strike at time zero and returns to the next heel strike at 100. By comparing the gait phase with the EMG, you can see when individual muscles are active in the gait cycle. In this application, the main goal is to learn *the timing sequence of* the major muscles during a complete gait cycle. For more details, consult the Appendix on EMG.

While the sEMG is useful for registering muscle activity in laboratory conditions, its accuracy and resolution are low, the electrodes are messy, and prolonged contact can cause skin problems such as pressure sores on the limb. Moreover, the myographic signal is essentially Gaussian noise, which requires complex signal processing and sophisticated classification algorithms for interpretation. Other sensing modalities for muscle activity are now available, as discussed further in Chap. 6.

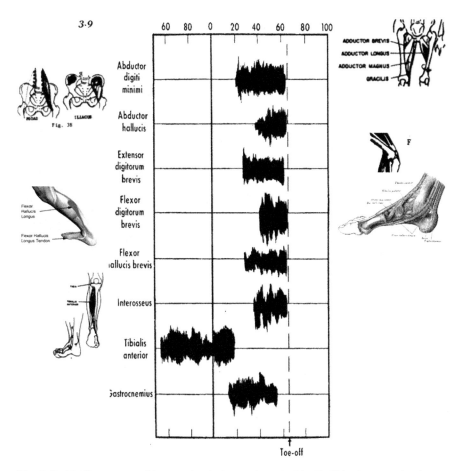

Fig. 3.17 EMG responses of leg muscles during a gait cycle. The EMG is shown as signal envelopes. J. Bone & Joint Surgery, 46-A-469, 1964. Fpnotebook.com

3.6.4 Measuring Locomotion

Goal: To learn the kinematics and dynamics of natural and prosthetic-assisted gait.

3.6.4.1 Introduction

As stated above, restoring lower limbs requires knowledge of the motions and forces that produce healthy gait. Systematical measurement of these variables at each joint and during a range of locomotory activities is done using the tools of gait analysis. Data from gait analyses of sound-limbed individuals have been catalogued and are used by prosthetic designers. With normative information, designers can achieve restorations that provide optimal comfort and efficiency of gait with the prosthesis. This chapter highlights a few aspects of locomotion, relative to prosthetic legs.

3.6.4.2 Phases of the Walking Cycle

Human gait can be described with simple parameters of cadence, stride, and balance, using only a stop watch and measuring tape. More precise information, i.e., joint angles and forces, can be obtained by photographically recoding subjects walking on a force plate, as noted above. Alternatively, walking subjects are fitted with IMUs that record all joint motions in all planes. A typical gait test involves subjects walking at a range of walking speeds, on a different terrains, such as slopes, textures, and steps. Evaluating subjects during their chosen walking speeds is advantageous, since this is natural behavior, not artificial gait alterations. With a force plate and camera, joint forces can be estimated from the ground reaction forces and kinematics. A complete gait analysis includes a record of motion and net moments in the sagittal plane at the hip, knee, and ankle joints.

Human locomotion is the transformation of coordinated angular motions occurring simultaneously at the various joints of the extremities (all four limbs), into forward progression of the body. All major joints interact with each other: knee-ankle, hip-knee, hip-pelvis, shoulder-elbow, and neck-head. The torso rotates about a vertical axis while oscillating laterally, while arms rotate in the sagittal plane. A thorough study of walking in the orthograde attitude would therefore include not only the influence of each of the limbs on the total displacement pattern but also a complete analysis of the action of major muscle groups. The following section summarizes the phases and joint activities during a gait cycle. The hip, knee, and ankle joints and their interactions are depicted.

All motions associated with gait represent the effects of muscular actions working both with and against gravity, which is a major factor in gait. Gravity is always present to varying degrees on Earth and ideally serves to energize and stabilize gait, conserve energy, and coordinate an efficient and graceful gait.

The stiffness, damping, and inertia of each limb are affected by both dynamic and kinematic factors. The dynamic factors include forces of both gravity and muscles. The kinematic factors include joint angular position and velocity.

The upright, bipedal walking cycle may be divided into two phases—the stance (or weight-bearing) phase and the swing phase. The **stance** phase of any leg (right leg in Fig. 3.18) begins at the instant the heel contacts the ground (initial contact) and ends at toe-off when ground contact is lost by the foot of the same leg. The **swing** phase begins as an acceleration at toe-off and ends at heel contact. The two feet are in simultaneous contact with the walking surface for approximately 25% of a complete two-step cycle, this part of the cycle being designated as the "double-support" phase. Two biometric measures of gait are "step length" and "stride length" as shown (Fig. 3.19).

3.6 Human Walking Kinetics

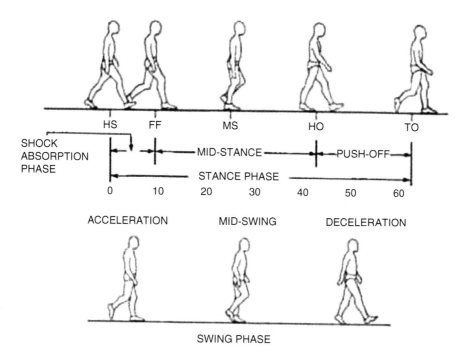

Fig. 3.18 Phases of gait. J. Hughes *, N. Jacobs *, O&P Library > POI > 1979, Vol 3, Num 1 > pp. 4–12 O&P Lib

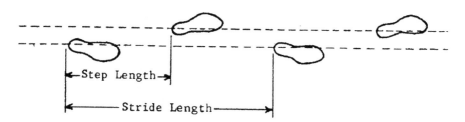

Fig. 3.19 Definitions of step and stride lengths. O&P Library

In reference to Fig. 3.18, and particularly to the curves in the region corresponding to the end of the swing phase (about 95% of a complete cycle), the knee joint reaches its maximum extension just prior to heel contact and that a period of knee flexion then initiated continues on into the stance phase. As seen in the muscle activity graphs, this decrease in the rate of knee extension at the end of the swing phase, while preparing for foot contact with the floor, is due primarily to the action of the hamstring muscle group, which is attached to the pelvis behind the hip joint and to the *tibia* and *fibula* below the knee joint. Tension in the hamstring group can cause both hip extension and knee flexion simultaneously.

Initial Contact Phase

Initial contact is normally the time when the heel (or other foot part if heel is missing or the foot is deformed) first contacts the ground. The label, "heel strike," is no longer used because it is invalid for those gait patterns in which the forefoot makes the initial contact, as exhibited by certain neurological disorders. Upon initial contact, *hamstring* contraction tends to bring the foot forcibly onto the floor, while the knee rapidly flexes. The impact force of heel strike is transmitted to and absorbed by both the ankle and the knee. The shock absorption allows a smooth descent of the body's center of mass. As the *hamstrings* gradually relax, the *quadriceps* quickly stiffen the knee joint, while the pretibial group stiffens the ankle joint. In this way the knee and ankle motions are coordinated, as the forefoot is smoothly lowered to the floor; shock absorption at footfall is provided by both knee and ankle actions.

Mid-stance Phase

Knee flexion during loading phase persists during mid-stance and reaches its maximum of approximately 20° during early mid-stance. As the body center of mass traverses over the knee, the upward thrust of the floor reaction moves forward toward the sole of the foot, thus gradually increasing dorsiflexion of the ankle and extending the knee. The knee requires stabilization from quadriceps during this phase. In this period, the ankle-knee interaction is controlling stability, with only minimal muscular activity in the groups acting about the hip and knee. The knee reaches a position of maximum extension about the time the heel leaves the ground, the calf group providing the resistance to knee extension and ankle dorsiflexion. As the heel leaves the ground, the knee again begins flexing, as the hip joint flexes. This sequence of controlled flexion at initial contact, release to allow gradual extension in mid-stance, and controlled flexion preparatory to swing accomplishes a smooth and energy-saving gait in sound-limbed persons.

Push-Off Phase

During the push-off phase, the knee is brought forward by action of the hip joint, and a sensitive balance is maintained by interaction of hip, knee, and ankle joints. The combined action has two purposes—to maintain the smooth forward progression of the body as a whole and to initiate the angular movements in the swing phase that follows. As the knee begins to flex (about the time the heel leaves the ground), the knee musculature must first resist the external effect of the force on the ball of the foot which passes through space on a line ahead of the knee joint. Then, as the knee is brought forward by hip joint action, so as to pass through and then anterior to the line of the force acting upward on the foot, the knee must reverse its action to provide controlled resistance to flexion by increasing *quadrice*ps activity. *Hamstring*

3.6 Human Walking Kinetics

activity in this phase would be antagonistic and may or may not occur, depending on the walking style.

The calf group continues to provide active plantar flexion during the entire push-off phase. At the time the toe leaves the floor, the knee has flexed 40–45° of the maximum of 65° it reaches during the swing phase. In sound-limbed persons, knee flexion in the swing phase is not due primarily to hamstring action, as might be supposed. Complete prosthetic restoration of normal function in the push-off phase is difficult, if not impossible. A proprioceptive sense of knee position by the amputee is necessary, as well as an active source of energy in the ankle. Because of the lack of an active source of ankle energy, initiation of knee flexion in amputees wearing a prosthesis must come from active hip flexion.

Swing Phase (*Quadriceps* Action)

The objective of swing phase is to get the foot from one position to the next smoothly while clearing the foot over any obstacles of terrain. At the start of the swing phase, the leg is flexed and is ready to use its potential energy to swing. Active extension of the ankle and flexion of the hip during the push-off phase. The knee is flexing and continues to flex after toe-off. During rapid walking this would result in excessive knee flexion and heel rise were it not for the action of the quadriceps group in limiting the angle of knee flexion to approximately 65° and then continuing to act to start knee extension. Knee extension continues as a result of a combination of pendulum effects owing both to muscle action and to the weight of the inclined shank and of the foot.

Work by the *quadriceps* is minimized during swing, due to gravity-driven hip flexion, acting as a pendulum as well as active hip flexion from the *iliopsoas*. These dynamics accelerate the knee forward and upward.

Mid-swing

Muscular activity is minimized during mid-swing as the leg swings like a pendulum with gravity. The leg can be modeled approximately as a simple pendulum, with knee locked during swing phase. During running, the leg behaves as a compound pendulum, with both hip and knee joints involved.

Terminal Deceleration (Hamstring Action)

At the end of the swing, the rate of knee extension must be slowed to decelerate the foot prior to heel contact. This "terminal deceleration" of the normal leg is accomplished by the *hamstrings*, which produce extension resistance at the knee.

3.7 Exercises

1. In reference to Fig. 3.15, the EMA is determined by the skeletal anatomy of the foot. How do foot metrics affect the efficiency of gait during push-off?
2. During gait, which leg muscle is least active? Which muscle activates prior to heel strike?
3. Write a flowchart describing a myoelectric control system with two degrees of freedom, starting with a raw EMG.
4. Describe what is meant by "signal envelope," and what is its purpose.
5. Describe a method for rectifying a signal, such as the EMG. How would this procedure affect the rms and mean of a signal, such as EMG?

Acknowledgment *Adapted from* Charles W. Radcliffe, M.S., M.E., the O&P Library.

Chapter 4
Lower-Limb Prostheses

Goal: To learn the key anatomy and available control mechanisms for restoring lower limbs.

4.1 Introduction

Because it is nearly impossible to walk following loss of a lower limb, the quest for limb replacements has been ongoing throughout history. As noted in Chap. 1, many solutions for limb loss have been developed over the ages, but up until the 1960s, leg amputees were fitted with wooden or metal sockets carved out by hand, not too dissimilar from the ancient products. These were rigid, heavy, and generally ill-fitting sockets that tended to slide up and down during gait, causing shear force injury to the residuum and the knee. Under these conditions, the residuum was subject to edema, i.e., accumulation of body fluids in the residuum, which made ambulation difficult. The introduction of the leather thigh-lace-up corset improved ambulation, since it carried the body weight safely and comfortably. Today, modern imaging and materials technology and better understanding of biomechanics in the developed world can provide comfortable restoration of mobility to persons with leg amputation.

The functional requirements of lower-limb prostheses are relatively simple compared with those of the upper limb, and thus modern LL prostheses are capable of restoring nearly natural gait. Normal human gait is accomplished by coordinated actions of muscles and gravity. For example, even the simplest phase of gait, leg swing, which is essentially pendular motion, requires programmed control of many joints. Since human gait is unique to each individual, and differs according to anatomy, physiognomy, and learned patterns, no two human beings walk in identical patterns. LL prosthetic technology is getting closer to reproducing the complexities of gait in various situations, by incorporating intelligent materials and devices, such

© Springer Nature Switzerland AG 2022
W. Craelius, *Prosthetic Designs for Restoring Human Limb Function*,
https://doi.org/10.1007/978-3-030-31077-6_4

as magneto-hydraulics, electric braking, and computer control, to better mimic natural human gait.

The two major categories of lower-limb prostheses are those for transfemoral Amputations (above the knee (AK)), and those for trans-tibial amputations, below the knee (BK). BK amputations are much more common than those of the AK, and more easily restored to near-normal ambulation. Persons with AK amputations require a prosthetic device with an artificial knee, while BK amputations will preserve function of the natural knee, as noted below. Both types of lower limb amputations require an artificial ankle.

4.2 Leg Functional Anatomy

The three major joints of the human leg, hip, knee, and ankle, are depicted in Fig. 4.1. The hip is essentially a ball and socket joint that rotates in all three axes: flexion-extension, adduction-abduction, and horizontal rotation; by combination of these rotations, the hip can also circumduct. The knee flexes and extends in one plane (sagittal) like a hinge, but it also slides while it rotates along the condyle. It operates as a "four-bar" mechanism, as shown in Fig. 4.1.

The *quadriceps* comprise four distinct component muscles that merge into one tendon inserting at the anterior portion of the tibia (Fig. 4.2). Thus, when the *quadriceps* tendon contracts, it rotates the tibia upward (knee extension), with an extension torque equal to its force magnitude times the lever arm between the tendon and the center of the femur condyle. The patella being embedded in the tendon thus plays at least two roles: supporting the *quadriceps* tendon between the femoral condyles and augmenting the lever arm of the *quadriceps* wrapped around the knee axis (Fig. 4.2). Thus torque is increased in proportion to the lever arm increase. The cartilaginous underbelly of the patella is smooth and lubricated, providing a slippery, frictionless cover as it glides over the anterior surface of the femur. With this anatomical arrangement, rotation of the shank during gait is aided by several passive components, as well as inertia, with little active muscular support. The shank thus moves much like a pendulum during walking.

4.2.1 Knee Function After BK Amputation

The most common amputation is "below knee" (BK). This surgery preserves knee joint function by leaving knee muscles attached to the tibia. While these muscle insertions can remain intact after simple BK amputation, some knee flexural ability can be lost, however, due to the inevitable loss of the *gastrocnemius* lever which originates from the posterior portion of each of the femoral condyles and inserts at the patella tendon (Figs. 4.3 and 4.4).

4.2 Leg Functional Anatomy

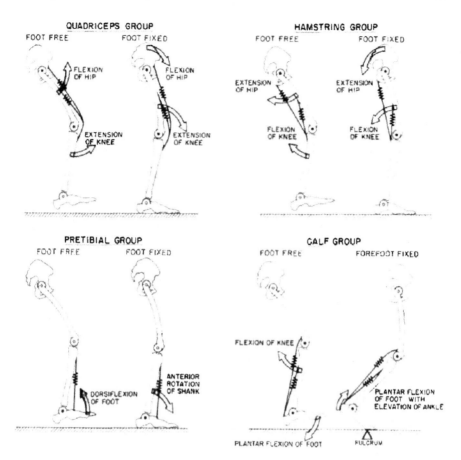

Fig. 4.1 Major joints, movements, and muscular actions of the normal lower extremity (schematic), showing the major mechanics in the parasagittal plane. Muscles are represented as springs. O&P Library

The knee joint itself, the tibial-femoral joint (TFJ), is shown in Fig. 4.4. The knee joint is comprised of the condyles (tops of), the femur, and tibia along with the patella. The anterior and posterior cruciate ligaments are both shown. Contraction of the *quadriceps* causes the patellar tendon to pull inward and upward, flexing the knee, as shown. A normal knee can rotate from 0° (extension, panel A) to beyond the 45° shown, up to 160 degrees of flexion in the sagittal plane, and can rotate a small amount in the x and y planes, and can slide in the x plane as well. During flexion, the humerus slides over the tibial condyle, through a combination of gliding and rolling actions, as depicted. Thus, the instantaneous center of rotation of the knee varies with the degree of flexion.

The patellar ligament is composed of tough but elastic fibers that stretch under tensile loads from knee flexion. Knee flexions also apply large compression in the anterio-posterior direction, i.e., the "y" axis of the body. *Quadriceps* contraction

Fig. 4.2 The muscles of the lower leg are depicted. The major knee flexors are the *Rectus Femoris* and *Vasti* muscles, labeled REC and VAS. These muscles comprise the *quadriceps*, which insert at the hip and reach down to the tibia and fibula, where they insert at points just below the knee joint. The gluteus muscles flex the hip, the hamstrings extend the knee, and the *gastrocnemius* and *soleus* flex the ankle. **Each muscle is represented as an active contractile component in series with a spring/dashpot. The *quadriceps* are the Rectus Femoris and the Vastus Lateralis, which extend the knee**

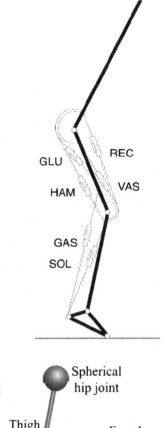

Fig. 4.3 *Artists view of leg muscles and knee joint.* https://doi.org/10.1007/s10846-019-01063-5. Jianfeng Li1 Shiping Zuo1 Chenghui Xu1 Leiyu Zhang1 Mingjie Dong1 Chunjing Tao2 & Run Ji2 Journal of Intelligent & Robotic Systems (2020) 98:525–538

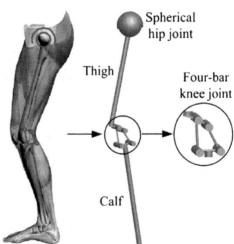

tightens its tendon to prevent relative motion between the patella and the tibia. Quadriceps contraction requires its tendon to withstand high compressive loads, and too much force will damage the tendon. The patella, however, in contrast, is a relatively soft gel and serves mainly as a lubricated gliding surface. Under most

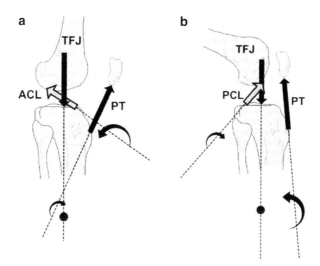

Fig. 4.4 Flexion of the knee joint by quadriceps, with contribution of the ACL (a) and PCL (b). This figure shows the patella acting as a passive spacer between the joint and the quadriceps tendon, which increases the moment arm, and hence torque, applied by the quadriceps on the tibiofemoral joint. This extra spacing can also assist femoral rotation, since tibial rotation otherwise will decline as the knee flexion angle increases. Another boost to knee flexion is the action of the ACL which helps extend the tibia. doi: 10.1371/journal.pone.0115670.g002, On the Role of the Patella, ACL and Joint Contact Forces in the Extension of the Knee. Daniel J. Cleather1*, Dominic F. L. Southgate2, Anthony M. J. Bull2

conditions knee extension is well controlled within safe limits during gait, climbing, jumping, and running.

To maximize residual function following amputation, the distal end of the *gastrocnemius* (Fig. 4.2) is surgically re-attached to the tibia, allowing the remaining musculature to assist the flexors and help stabilize the fibula with respect to the tibia. Without this reconnection, knee dislocation is likely, and active knee motion is lost. BK surgery thus restores almost completely the moment generated about the knee in the parasagittal plane. The mediolateral stability of the BK amputee usually is not affected, because the ligaments are generally not harmed in surgery.

A rendering of the muscles on an intact right leg, and the residual muscles of the left leg (amputated below the knee), is shown in Fig. 4.5. This sketch compares the machinery of the sound left limb and a BK prosthetic right limb during gait. Here the upper torso is shown as a solid cylinder. Note that the prosthetic ankle is driven by a motor, which attempts to replace the function of the *gastrocnemius, soleus*, and *tibialis* muscles.

The relative moments at the prosthetic knee during four phases of gait during stance are shown in Fig. 4.6. Note there is a brief extension moment at heel strike, followed by a bell-shaped (increasing and decreasing) moment during weight acceptance, and then a large extension moment during terminal stance.

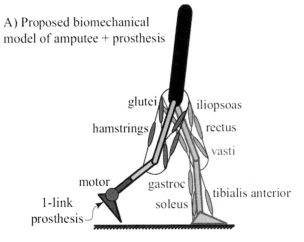

Fig. 4.5 (doi: 10.1371/journal.pone.0115670.g004, Daniel J. Cleather1*, Dominic F. L. Southgate2, Anthony M. J. Bull2)

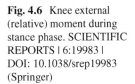

Fig. 4.6 Knee external (relative) moment during stance phase. SCIENTIFIC REPORTS | 6:19983 | DOI: 10.1038/srep19983 (Springer)

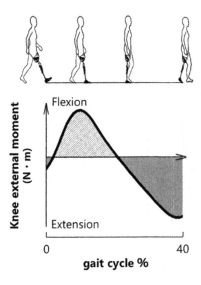

4.3 Natural and Prosthetic Knee Motion During a Gait Cycle

Kinematic actions of the human knee during a complete gait cycle (from right foot contact to next right foot contact) are illustrated in Fig. 4.7. The muscular extension torques applied during stance phase are depicted as springs on the posterior aspect of the knee. These torques are needed to counterbalance the natural flexion torques occurring at stance phase.

Early designs of prosthetic knees were a simple hinge joint. Today, prosthetic knees can be controlled much like the natural knee, using sensors and actuators, that control both stance and swing phases using pneumatic, hydraulic, or electronic systems that attempt to mimic the muscles. As seen in Fig. 4.8, the condyles are

4.3 Natural and Prosthetic Knee Motion During a Gait Cycle

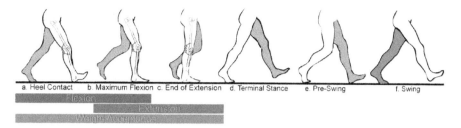

Fig. 4.7 A complete gait cycle showing knee function

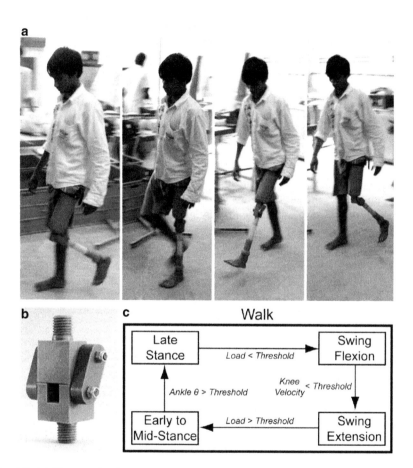

Fig. 4.8 (a) Photograph of gait cycle with a Jaipur single-axis prosthetic knee. SAMUEL R. HAMNER, VINESH G. NARAYAN, and KRISTA M. DONALDSON Annals of Biomedical Engineering, Vol. 41, No. 9, September 2013 (_ 2013) pp. 1851–1859, DOI: 10.1007/s10439-013-0792-8 Springer. PLoS ONE 13(1): e0191310. https://doi.org/10.1371/journal.pone.0191310. (**b**) Finite state machine for level ground walking. doi: 10.1371/journal.pone.0099387.g001. (**c**) Control logic for the Jaipur prosthesis

modeled as a cylinder around which rotate a top and bottom element, representing the femur and tibia, with muscles modeled as stiff connectors. Logical control of the Jaipur knee is depicted in Fig. 4.8c. Muscles are represented as springs and motors acting antagonistically, similar to the muscular control of the natural knee.

4.4 Prostheses for the AK Case

Prior to the 1960s, the most common surgery for vascular disease was above-knee (AK) amputation. This procedure is more drastic and limits ambulation, compared with BK amputation, since it requires a replacement knee. Advanced artificial knees can control their movement, similarly to natural gait, as described below. The simplest knee controller is a high friction joint, with springs to prevent collapse. Human knee function can be mimicked more closely with advanced designs, such as the four-bar knee (Fig. 4.9), the hydraulic leg (Fig. 4.10), the Rheo knee (Figs. 4.11, 4.12 and 4.13), and the C-Leg (Fig. 4.14).

Lower-limb prostheses have been advancing rapidly since 1980 as depicted in Fig. 4.15. Note that some advanced prostheses were designed and available on a limited basis prior to 1980.

The modern above-knee leg prosthesis, like the natural leg, represents a compound pendulum, with actuated thigh, shank activated at the knee, and a functional ankle, both of which are under control by gravity. The two members of the compound pendulum are the thigh and shank, each of which rotate about the knee joint.

 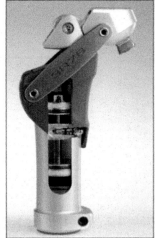

Fig. 4.9 Four-bar hydraulic knee. https://www.google.com/url?sa=i&source=images&cd=&ved= 2ahUKEwjKpaDK57XmAhWMm-KHc9PAlUQjB16BAgBEAM&url=https%3A%2F%2Fwww. medicaldesignbriefs.com%2Fcomponent%2Fcontent%2Farticle%2Fmdb%2Ffeatures%2Farticle s%2F20250&psig=AOvVaw0tVZGyUggk2Fzx49stV_K3&ust=1576435876612368

4.4 Prostheses for the AK Case

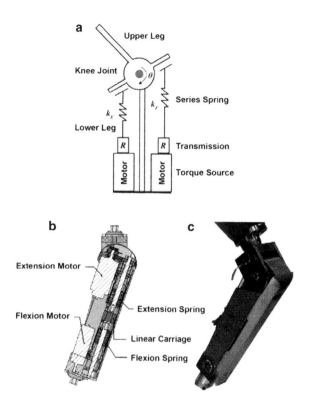

Fig. 4.10 Design of motorized mimic of human knee

The simplest knee replacement consists of a high friction joint, with springs controlling extension and flexion in order to prevent collapse. One such prosthetic knee is the single axis Jaipur knee (Fig. 4.8a, b), in which flexion and extension occur around a single axis, like a hinge joint. Advantages of the single axis knee are its reliability, simplicity, low maintenance, and low cost. This simple prosthetic knee is used worldwide.

Figure 4.15 shows a single gait cycle by a trans-humeral amputee wearing the Jaipur knee. The knee bends at toe-off shown in the right panel. Second panel shows the straightened knee during leg swing, right before heel strike. Third panel shows stance phase, with knee locked, and fourth panel shows the beginning of the next cycle, right after toe-off. This simple knee controls its friction during the critical phases of gait.

The logic built into the Jaipur knee is shown in Fig. 4.15b. This is known as a "finite state machine." At the late stance phase, pictured at the right panel, the swing phase begins when the load on the foot becomes less than a pre-assigned threshold. Thus the logic determines at what point the knee should loosen, and begin to swing. Next, when the swinging leg velocity slows, and becomes below threshold, the knee begins to extend. Next when the load exceeds threshold, this represents early to mid-stance, when the body weight is transferred to the ground, and then the sequence repeats.

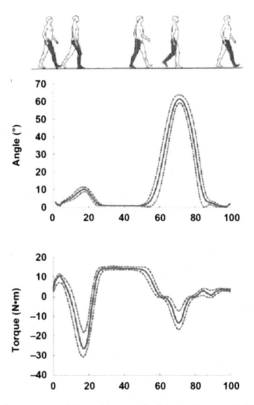

Fig. 4.11 Results of amputee walking with motorized knee. Ernesto C. Martinez-Villalpando, SM;1 Hugh Herr, PhD1–2*Agonist-antagonist active knee prosthesis: A preliminary study in level-ground walking JRRD Volume 46, Number 3, 2009 Public Domain. DOI:10.1682/JRRD.2008.09.0131. Freely available

Fig. 4.12 Representation of knee with flexor and extensor muscles represented as springs

4.5 Mimicking the Human Lower Limb

Advanced prosthetic knees which closely mimic the anatomy of the human knee have been designed; a structural mimic is the four-bar knee, Fig. 4.9. (Compare with Fig. 4.1.)

4.5 Mimicking the Human Lower Limb

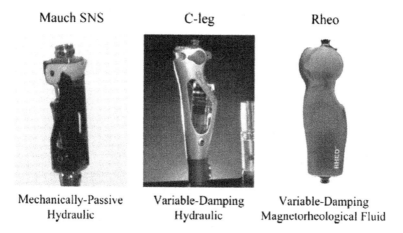

Fig. 4.13 Modern artificial knees. **License Number** 4863161298260 Am. J. Phys. Med. Rehabil. ● Vol. 84, No. 8 Johansson JL, Sherrill DM, Riley PO, Bonato P, Herr H

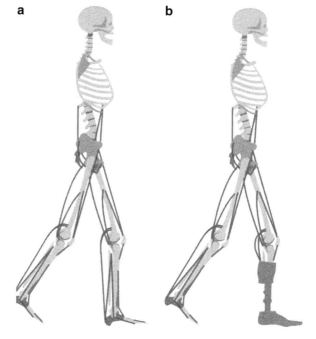

Fig. 4.14 Muscles used in sound-limbed gait, and after below-knee amputation. (**a**) Pre-limb loss, (**b**) post-limb loss. Note the Anterior tibialis and gastrocnemius can remain partially functional for those persons with BK amputation. Elizabeth Russell Esposito1,2, Ross H. Miller3,4*PLoS ONE 13(1): e0191310. https://doi.org/10.1371/journal.pone.0191310

Another design attempts to mimic the agonist-antagonist actions of the *quadriceps* and *hamstrings*, as shown in Fig. 4.10. The knee joint is hinge-type, whose angle is controlled by reciprocal springs and dampers for compliance. In order to mimic muscular control of the knee, the motors are controlled by events and parameters measured during the gait cycle, similarly to that used by natural muscles. Thus, the knee stiffens at heel strike and loosens during swing phase.

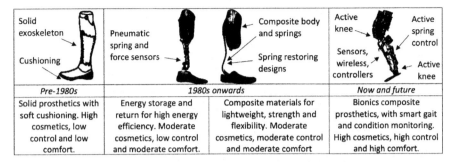

Pre-1980s	1980s onwards		Now and future
Solid prosthetics with soft cushioning. High cosmetics, low control and low comfort.	Energy storage and return for high energy efficiency. Moderate cosmetics, low control and moderate comfort.	Composite materials for lightweight, strength and flexibility. Moderate cosmetics, moderate control and moderate comfort	Bionics composite prosthetics, with smart gait and condition monitoring. High cosmetics, high control and high comfort.

Fig. 4.15 Timeline of lower-limb prosthetic developments. Energies, freely available Yu Jia 1,2,*, Xueyong Wei 3, Jie Pu 2, Pengheng Xie 2, Tao Wen 2, Congsi Wang 4, Peiyuan Lian 4, Song Xue 4 and Yu Shi 2

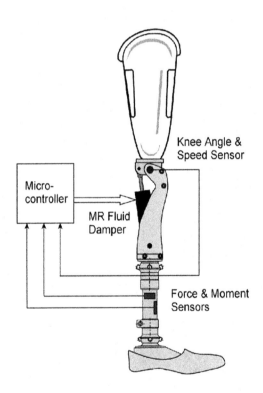

Fig. 4.16 Rheo knee: sensors continuously measure the state of the knee through the angle, velocity, axial force, and bending moment. A controller receives these sensory signals and delivers electrical current to proportionally adjust knee stiffness, by damping during foot strikes, reducing the chance for knee collapse. J. David Carlson, Lord Corporation, Materials Division, Cary, North Carolina, USA

4.5 Mimicking the Human Lower Limb

Results of walking trials with the motorized knee are shown in Fig. 4.11. As the heel touches the ground there is a small flexion, followed by straight knee during stance, then a large sinusoidal-like flexion. Maximum flexion torques on average are about 25 N-M. Note the multiple traces are average and standard deviations based on ten walking trials. The joint torque produced by the ankle muscular apparatus is a function of spring torsion stiffness and angle, simply estimated as $\tau = -k.q$, where τ is the joint torque as a function of time, k is the torsional stiffness, q is the angle of the ankle as a function of time.

Three advanced knees, the Mauch, C-Leg, and Rheo, are shown in Figure 4.13. Figure 4.14 illustrates how BK amputees may retain control of their shank. Details of operation of the Rheo knee are illustrated in Fig. 4.16.

Evolution of modern lower-limb prostheses is illustrated in Fig. 4.15.

Future improvements of this knee are planned to include nonlinear springs that better mimic muscle action as shown in Fig. 4.12.

Several other active knees are available, the most common of which are shown in Fig. 4.11.

4.5.1 Gait Muscles Summarized

To summarize the foregoing treatment of artificial legs, the design goals allow for maximal use of the natural muscles, while providing artificial replacements of reflexes involved in stable gait. For example, Fig. 4.14 shows how some functional control can be retained with proper design. In the sound-limbed model (left) all major joint muscles are depicted, including the abdominal, lumbar, hip, knee, and ankle. The anterior and posterior muscles of the shank can remain operational. The ankle muscles, however, including the *gastrocnemius, tibialis anterior, soleus, extensor digitorum, and flexor digitorum* are absent, and their function must be replaced artificially.

4.5.1.1 Artificial Knee Control

During each gait cycle the muscles of the natural knee joint execute a patterned activation sequence to modulate joint friction. For example, knee friction needs to be high at heel strike, so both the *quadriceps* and hamstrings tighten; during swing phase, they relax. This pattern is responsible for graceful walking. Corresponding muscle actions are tightening at heel strike, and loosening during swing. In the absence of this multi-muscular coordination, friction is not controlled, and the knee joint is free to rotate when any force is applied, leading to a fall. Combined with the effects of inertia and gravity during gait, the shank would rotate wildly and slam into extension as it rotates forward, except at a very slow rate of walking. Restoring the same kind of coordinated control in an artificial knee joint requires a control mechanism coordinated with the gait phase. Modern technology is capable of orchestrating this control, which can restore near-normal walking, and even running, using various mechanical friction controls about the knee joint. To restore near-natural gait, a variety of modern knees use many types of frictional control.

While externally powered knees as described above can restore near-natural gait under ideal conditions, their reliability is not as high as the strictly mechanical limbs. Battery failure is a particularly troubling occurrence, which can lead to functional shutdown, and subsequent stumbling or worse. While "fail-safe" mechanisms are incorporated in some of these devices, they are still not immune to catastrophic failure leading to falls. One common safety design feature deployed in all types of leg prostheses is aligning the knee axis behind those of the hip and ankle. While this fix is not ideal, since it is somewhat unnatural, it can stabilize the knee joint in extension, which discourages sudden flexion, and subsequent fall. Still, walking on any artificial leg in the varied environments, uneven surfaces, stairs, and slopes is a challenge requiring particular attention by the user. This dilemma, presented by ambulating in the real world, highlights the need for better, more natural control of leg prostheses.

4.5.1.2 Magnetorheological Knee

An advanced design for knee control is the magnetorheological fluid (MRF) knee, shown in Fig. 4.16. This knee exploits the feature of electro-rheological fluid (ERF) that has been developed for controlling automobile transmissions, brakes, and other moving components. ERF viscosity can be controlled by electrical currents, and thus can provide variable mechanical resistance to mechanisms, such as knee joints. The principle of operation of ERF is shown in Fig. 4.17. Figure 4.17a depicts an ERF containing randomly situated magnetic particles. Figure 4.17b shows an electrical field causing the particles to line up, increasing the fluid viscosity, which then increases friction in a hydraulic mechanism, such as a brake, or a prosthetic (or orthotic) knee. Knee friction can thus be controlled in two phases: stance phase—friction high and swing phase—friction low.

The embodiment of a MRF knee is shown in Fig. 4.18 and the design logic is shown in Fig. 4.20.

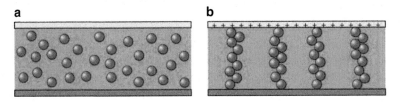

Fig. 4.17 Electrical fields increase the viscosity of electro-rheological fluids by aligning the dissolved particles. (**a**) No electrical field. (**b**) With electrical field. Figure uploaded by *Gaweł Żyła to* Research Gate. Content may be subject to copyright. J. David Carlson, Lord Corporation, Materials Division, Cary, North Carolina, USA

4.5 Mimicking the Human Lower Limb

The logic of the MRF knee is diagrammed (Fig. 4.18). *Another advanced knee is based on a hydraulic cylinder to control knee stiffness according to gait phase*; Otto Bock C-Leg Knee control system (Fig. 4.19), and control modes for the C-Leg are diagrammed in Fig. 4.20.

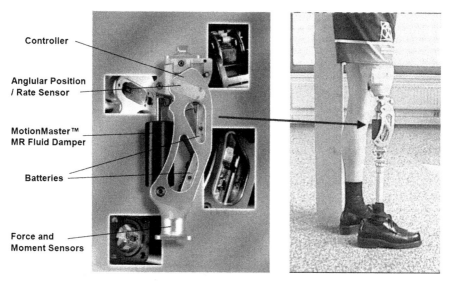

Fig. 4.18 The magneto-rheo knee works well going down stairs, one of the most difficult maneuvers for AK amputees. Smart Prosthetics Based On Magnetorheological Fluids. J. David Carlson*a, Wilfried Matthis**b and James R. Toscano*a. Lord Corporation, Materials Division, b Biedermann Motech GmbH

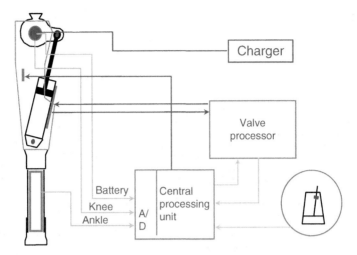

Fig. 4.19 Schematic of the control process of a hydraulic knee. Otto-Bock.com (Permission pending)

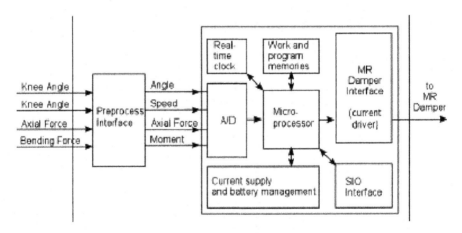

Fig. 4.20 Control of the MRF Knee, Lord Corporation, Materials Division, b Biedermann Motech GmbH

4.6 Hip Disarticulation and Bilateral Leg Cases

4.6.1 Prostheses for Hip Disarticulation and Hemi-pelvectomy Cases

Hip disarticulation is total loss of the leg, which requires a complete replacement from the foot to the hip. The leg socket must be attached to the residual pelvis with straps and aligned for maximal safety and comfort during standing. The hip joint

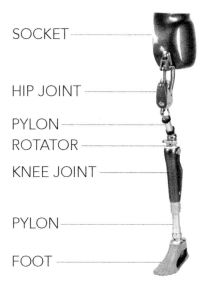

Fig. 4.21 Hip disarticulation socket with hip joint. http://www.pandocare.com/hemipelvectomy-hip-disarticulation/

must carry the weight of the limbs and joints and allow for ambulation. The joints must be particularly designed for safety, with high friction or controlled hydraulics to prevent falling. An advanced socket for hip disarticulation is shown in Fig. 4.21.

Training persons with hip disarticulation follows the same pattern as that for above-knee cases, but as would be expected, these cases pose special difficulties. Generally, at least one cane is required by those with bilateral loss or hip disarticulation.

4.7 Restoring Ambulation

The goal in training the leg amputee is to enable him/her to walk efficiently and gracefully. As modern technologies, including imaging and digital communication with mobile devices, are improving, the learning curve for using prostheses is improving considerably, and instruction is more accessible to users. The type and intensity of training can begin as soon as the client is provided with a comfortable prosthesis. Goals of training are to teach the patient proper methods of donning the prosthesis, caring for the residuum, arising after a fall, and use of canes and crutches when necessary. The ability to balance is the first prerequisite in learning to walk, so balancing is the first lesson for new amputees. When rhythmic gait has been accomplished, more difficult tasks are learned, such as pivoting, turning, negotiating stairs and ramps, and sitting on and arising from the floor. Many adjustments need to be made before the prosthesis will feel comfortable, and the user will feel confident.

The amputee's adjustment to the new limb depends on many factors, especially the level of comfort provided by the socket, as outlined in Chap. 9. The prosthesis transmits large forces at heel strike, which transmit to the entire body, including the intact skeletal structures and muscles. The sound limb may experience overloading, since it attempts to compensate for the lack of prosthetic support. A large part of the adaptation process can be characterized as developing "agency" with the new appendage.

While most BK amputees retain much of their leg musculature intact, persons with transfemoral (AK) amputation typically undergo severe muscle atrophy of their entire residual limb, losing much of their muscle power despite the almost continuous activation of the hip muscles during gait. Hip muscles are needed to maintain the stump in the socket and to stabilize the pelvis. Thus there is a need to provide auxiliary force to the hip during swing phase.

4.8 Exercises

1. Give and illustrate an example of reciprocal joint control by muscles. Label the agonist and antagonist muscle and describe how they work together for smooth motions.
2. Sketch a knee model showing its range of motion.

Acknowledgment The *OandP* Library *(Adapted from* A. Bennett Wilson, Jr., B.S.M.E.).

Bibliography

1. Wong JD, Bobbert MF, van Soest AJ, Gribble PL, Kistemaker DA. https://doi.org/10.1371/journal.pone.0150019
2. Cleather DJ, Southgate DFL, Bull AMJ. https://doi.org/10.1371/journal.pone.0115670.g004
3. Sci Rep. 6:19983. https://doi.org/10.1038/srep19983 (Springer).
4. Jia Y, Wei X, Pu J, Xie P, Wen T, Wang C, Lian P, Xue S, Shi Y. Energies, freely available.
5. Hamner SR, Narayan VG, Donaldson KM. The Jaipur knee. Ann Biomed Eng. 2013;41(9):1851–9. https://doi.org/10.1007/s10439-013-0792-8. Springer.
6. https://www.google.com/url?sa=i&source=images&cd=&ved=2ahUKEwjKpaDK57XmAhWMm-Hc9PAlUQjB16BAgBEAM&url=https%3A%2F%2Fwww.medicaldesignbriefs.com%2Fcomponent%2Fcontent%2Farticle%2Fmdb%2Ffeatures%2Farticles%2F20250&psig=AOvVaw0tVZGyUggk2Fzx49stV_K3&ust=1576435876612368.
7. Martinez-Villalpando EC, Herr H. Agonist-antagonist active knee prosthesis: a preliminary study in level-ground walking. JRRD. 2009;46(3):361–74. Public Domain.
8. https://doi.org/10.1682/JRRD.2008.09.0131. Freely available.
9. Esposito E, R, Miller RH. PLos One. 13(1):e0191310. https://doi.org/10.1371/journal.pone.0191310.
10. Johansson JL, Sherrill DM, Riley PO, Bonato P, Herr H. Am J Phys Med Rehabil. 2005;84(8):563–75.

11. Carlson JD. Lord Corporation, Materials Division, Cary, NC, USA.
12. Same.
13. Carlson JD, Matthis W, Toscano JR.
14. Lord Corporation, Materials Division, b Biedermann Motech GmbH.
15. Otto-Bock.com.
16. http://www.pandocare.com/hemipelvectomy-hip-disarticulation/.

Chapter 5
Upper Limb Function

Goal: Learn the mechanisms of the human upper limb.

5.1 Introduction

The human upper limb can perform an almost infinite variety of tasks, ranging from the most intricate and graceful to the most forceful. Employing 24 muscle groups, each controlled by many motor and sensory nerves, the hand is endowed with great dexterity. In comparison with lower-limb (LL) function, restoring upper-limb (UL) function is far more difficult. As we have seen, legs mainly need to bear weight comfortably and perform just a few different motions repetitively, using only 2 degrees of freedom. In contrast, the UL employs at least 28 degrees of freedom, with the hand accounting for 24 degrees. Thus the UL can perform a limitless number of motions and functions, only some of which can be produced by prostheses.

Artificially reproducing even a fraction of human hand functions is a daunting task. As we will see in Chap. 9, restoring UL function primarily requires improving the human-machine interface (HMI). The HMI must not only translate human volitional commands into the desired movement and force but also must sense and control the resulting response of the prosthesis. Presently, the HMI is the weak link in the chain of prosthetic restoration of natural limb function, and a better HMI requires a more complete understanding of brain-limb communication, as discussed in Chap. 6.

5.2 Human Hand Function

Much of hand functionality depends on (1) the ability of the arm to position itself properly and (2) the brain to communicate the appropriate volitional signals to the hand. Having evolved and grown together, the hand and brain form a symbiotic relationship. Hand functionality is critical to early human development, especially of brain functions, whose circuitry forms in synergy with manipulation. Our hands are useful not only for manipulation but also for communication, thinking, and expressing emotion. Here, we outline the basic structures and operation of the human upper limb, as they relate to its artificial restoration.

5.3 Functional Anatomy, Motions, and Dynamics of the UL

The major axes of the UL joints are shown in Fig. 5.1. Note that the shoulder has three axes: adduction/abduction θ_1, flexion/extension θ_2, and rotation θ_3, (external/internal); elbow flexion/extension motions, θ_4; the wrist has three degrees of freedom: flexion/extension θ_5, radial/ulnar deviation θ_6, and pronation/supination θ_7. Thus, the arm, including wrist, has seven degrees of freedom (DoFs).

The shoulder is a ball-and-socket joint with three DoFs as shown in Fig. 5.1a, b. The effective number of arm DoFs is actually more when the movements of the torso are considered. Figure 5.1b shows the rotational axes of upper and lower arms along x, y, and z axes. Figure 5.1c shows the workspace range of the UL in a horizontal plane. Anatomical terminology for the UL is depicted in Fig. 5.2a. Muscles of the arm are depicted in Figs. 5.2b, c, 5.3, and 5.4.

When combining motions of its three major joints, the arm can execute additional motions as shown in Figs. 5.5 and 5.6.

5.4 The Hand

The volar aspect of the palm and digits, shown in Figs. 5.7 and 5.8, is covered with copious subcutaneous fat and a relatively thick and folded skin that permits easy bending and grasping in prehension.

The fat pads underlying the volar skin are tightly bound down to the skeletal elements; this arrangement resembles that of quilted pillows. This feature allows for steady non-slip gripping of objects.

The three *sulci*, or furrows at each finger, allow for easy bending of the fingers and tend to capture the pockets of subcutaneous fat in between. In comparison with artificial hands, the human hand is much more efficient in design and requires much less energy and torque for grasping. For this reason, cosmetic gloves for hand prosthesis are rejected by many users.

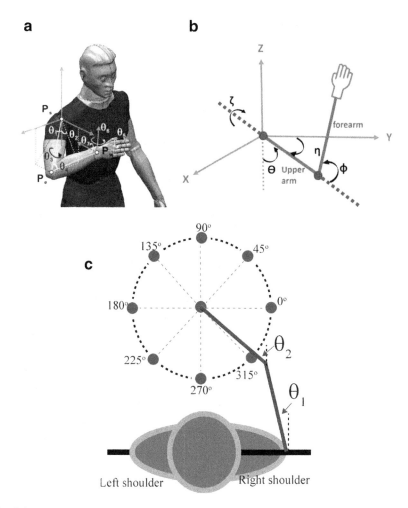

Fig. 5.1 (a) Axes and degrees of freedom of the human upper limb. Exp Brain Res (2017) 235:1627–1642 DOI 10.1007/s00221-017-4890-y. Zhi Li1 · Dejan Milutinović2 · Jacob Rosen Springer. (b) Axes and DoFs (angles) of human arm. (c) Workspace of typical arm in horizontal plane

5.5 Common Upper-limb Positions and Forces

The resting human hand natuarally assumes a characteristic posture, hanging loosely at the side. The resting wrist takes a mid-position in which, with respect to the extended forearm axis, it is dorsiflexed 35° (Fig. 5.9 **top**). This angle happens to be the position of greatest prehensile force, as seen in Fig. 5.10, **bottom**. The midposition for radial or ulnar flexion allows the MCP joint center of digit III to lie in the extended sagittal plane of the wrist.

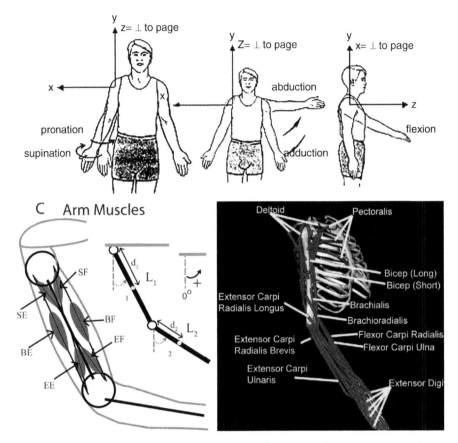

Fig. 5.2 (**a**) Anatomical positions and motions of the arm (O&P Library). (**b**) Muscular control of human upper arm, showing centers of mass of upper and lower segments. *SE/SF* shoulder extensor/flexor, *BE/BF* elbow extensor/flexor. https://doi.org/10.1371/journal.pone.0179288.g001. (**c**) Muscular control of entire human arm

Pinching and palmar prehension are fundamental tasks necessary to accomplish the activities of daily living. The conformations of fingers and thumb during these tasks are shown in Fig. 5.11.

The thumb is essential to prehension (Fig. 5.12). The versatility of the thumb lies in its carpometacarpal joint, which allows rotations and flexions, as well as flexion/extension in the carpal joint. The thumb can thereby position itself in three dimensions, while performing a variety of flexion/extension patterns within a spherical space, in contrast to that of the other four digits, which mostly operate in a plane. The thumb can oppose all the other digits, as well. This versatility is difficult to reproduce artificially.

Another versatile feature of the upper limb is its ability to rotate the forearm (with the hand) nearly 180°. When fully extended, the total rotation of the hand reaches almost 360° in healthy individuals, as shown in Fig. 5.13.

5.5 Common Upper-limb Positions and Forces

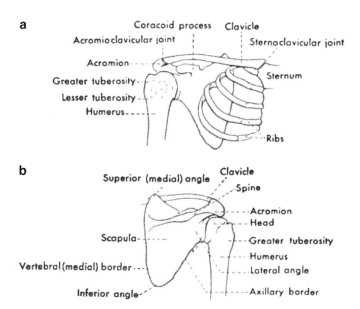

Fig. 5.3 Anatomy of the shoulder region. (**a**) Anterior view and (**b**) posterior view of the skeletal region. CRAIG L. TAYLOR, Ph.D., O&P Library 1955

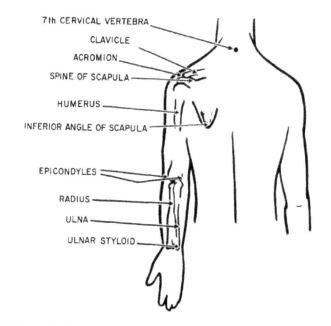

Fig. 5.4 The bones distal to the shoulder

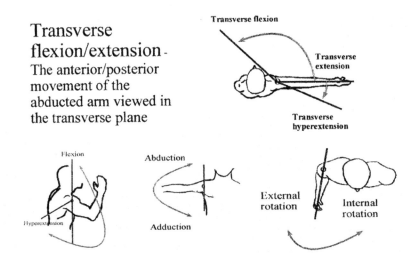

Fig. 5.5 Major bones of a (left) limb. CRAIG L. TAYLOR, Ph.D., O&P Library 1955

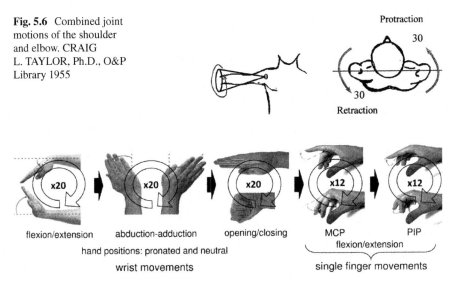

Fig. 5.6 Combined joint motions of the shoulder and elbow. CRAIG L. TAYLOR, Ph.D., O&P Library 1955

Fig. 5.7 Typical hand and wrist motions. Wrist and index, middle, ring, and little fingers metacarpophalangeal (MCP) and of the proximal interphalangeal (PIP) joint perform. Numbers represent repetitions of the movements. doi:10.1371/journal.pone.0109943.g001. Marco Gazzoni1*, Nicolò Celadon1,2, Davide Mastrapasqua1, Marco Paleari2, Valentina Margaria2, Paolo Ariano2*

The exquisite precision by which hands manipulate objects of any size, weight, stiffness, slipperiness, temperature, fragility, and shape features is a function of the human sensorimotor system, which can sense and react to any of these features for a secure grasp.

5.5 Common Upper-limb Positions and Forces

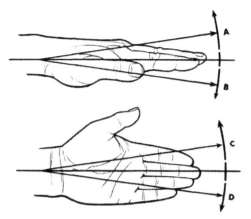

Fig. 5.8 Wrist flexion/extension and radial/ulnar deviation. The Palmar Digital Pads are depicted. The small ridges on the figures are known as Monticuli. **CRAIG L TAYLOR, Ph.D.,1 AND ROBERT J. SCHWARZ, M.D., O&P Library.** doi:10.1371/journal.pone.0109943.g001 Marco Gazzoni1*, Nicolò Celadon1,2, Davide Mastrapasqua1, Marco Paleari2, Valentina Margaria2, Paolo Ariano2

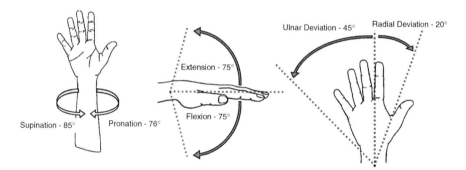

Fig. 5.9 Hand/wrist positions and prehension **forces. CRAIG L TAYLOR, Ph.D.,1 AND ROBERT J. SCHWARZ, M.D., O&P Library.** PLOS ONE, | https://doi.org/10.1371/journal.pone.0203861, September 12, 2018. **Wondimu W. Teka1*, Khaldoun C. Hamade2, William H. Barnett3, Taegyo Kim2, Sergey N. Markin2, Ilya A. Rybak2, Yaroslav I. Molkov3.,** https://doi.org/10.1371/journal.pone.0179288, June 20, 2017. DoFs and RoMs of typical human wrist. Neil M. Bajaj, Student Member, IEEE, Adam J. Spiers, Member, IEEE, and Aaron M. Dollar, Senior Member, IEEE DOI 10.1109/TRO.2018.2865890

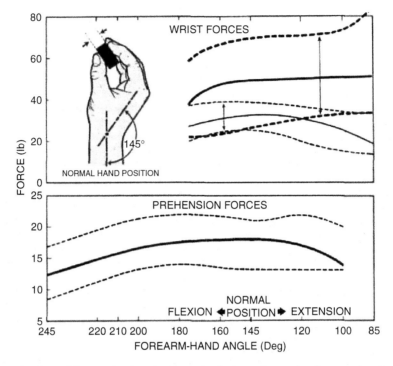

Fig. 5.10 Effect of forearm-hand angle upon wrist flexion and extension forces and prehension forces. The top graph is the relationship between the angle between the hand and forearm and wrist forces. The dark solid line represents the average forces in flexion and the light solid line is the average force in extension. The dotted lines represent the standard deviation. The bottom graph is the relationship between the forearm-hand angle and prehension forces. **CRAIG L TAYLOR, Ph.D.,1 AND ROBERT J. SCHWARZ, M.D., O&P Library**

5.6 Hand Control

Controlling the large number of muscles that drive the human hand is a large area of the cerebral cortex, whose size approximates the sum total area devoted to arms, trunk, and legs combined. In addition to the strictly motor portion of the brain, there are large sensory areas devoted to hand sensing. These large areas, containing millions of neuronal circuits, enable the hand to learn and perform its amazing feats of dexterity.

The hand itself is endowed with a dense network of sensory organs, enabling the fine tactile and kinesthetic sensitivity we enjoy. The touch threshold in the fingertip, for example, is 2 g per sq. mm, as compared to *33* and 26 g per sq. mm for the forearm and abdomen, respectively. The three major types of movement are (1) *fixation* accompanied by muscle co-contractions; (2) directional movements ranging from slow to rapid with controlled intensity, and rate; and (3) *ballistic* movements.

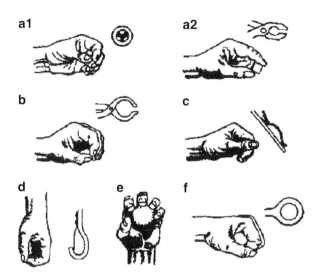

Fig. 5.11 Grasping positions. Flexor groups, inserted distally on the arm, produce cylindrical-grasp prehension by "rolling up" the hand in a "3-jaw chuck" configuration. Types of **Prehension:** (**a1**) palmar prehension (three-jaw chuck), (**a2**) palmar prehension (two finger), (**b**) tip prehension, (**c**) lateral prehension, (**d**) hook prehension, (**e**) spherical prehension, (**f**) cylindrical prehension. **O&P Library**

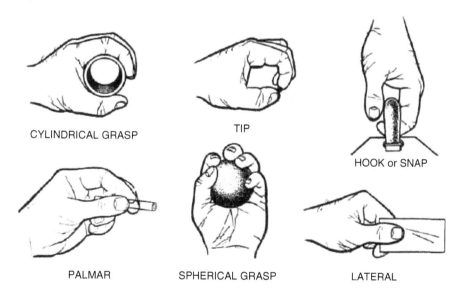

Fig. 5.12 Common hand functions. Keller, Taylor, and Zahn (1947): DESIGN OF ARTIFICIAL ARMS AND HANDS FOR PROSTHETIC APPLICATIONS. Weir, F. ff STANDARD HANDBOOK OF BIOMEDICAL ENGINEERING AND DESIGN

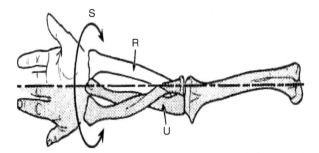

Fig. 5.13 Forearm rotational design. If the arm is fully extended the hand can rotate almost 360°. Due to the contributions of the shoulder and elbow. *U* ulna, *R* radius, *P* pronation, *S* supination. **CRAIG L TAYLOR, Ph.D.,1 AND ROBERT J. SCHWARZ, M.D. O&P Library**

5.7 Manipulation

Movements of any joint or appendage are the result of a net force coming from several muscles including both agonist and antagonist acting together. The balance of agonist/antagonist forces depends on the speed, intensity, and rate of the movement, and there is always some degree of co-contraction, to stabilize the motion and allow smooth changes in force and velocity. Examples of muscular activities range from writing, self-care, and playing music to forceful activities such as sports and manual labor. Included are actions involving differential and integrated motions of the digits. With intensive practice, significant increases in the facility of manipulation, even with simple operations, may be achieved, although individuals differ markedly in dexterity. The average individual has latent potential for skill development. Dexterity does not correlate well with mental ability or hand anatomy.

Postures assumed by a person's hand, wrist, and fingers are highly dependent on the tasks and environment of the user. Modern digital technology has markedly changed the usage patterns of hands, and new designs are needed to accommodate these changes. For example, the extended and rigid index finger is no longer needed for telephone use; however considerably more subtle features are required instead.

5.8 Arm Actuators

The forces and torques operating at the natural elbow are shown in Fig. 5.14. Elbow torque is a function of elbow angle, as shown. Maximum elbow flexion occurs near 80°. The elbow mainly performs flexion and extension and plays a role in supination/pronation as shown.

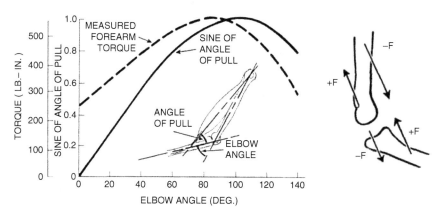

Fig. 5.14 Torques experienced at different elbow angles. The elbow joint experiences reciprocal force-couples (agonist/antagonist), pictured on the right. O&P Library

5.9 UL Joint Torques

Each joint is activated by their specific tendon, which connects specific muscles to the joint. For example, Fig. 5.14 shows the biceps muscle and its tendon pulling on the lever arm at the elbow. The torque at the joint is proportional to the tendon force times the lever arm. Since the *biceps* is a large muscle, and the lever arm is relatively long, the joint torque can be relatively large, exceeding 500 in. lb, as shown. The applied torque is roughly proportional to the sine of elbow angle, which relates directly to the lever arm. Thus the torque peaks at roughly mid-flexion. Hand and finger torques are similarly produced along the axis of the muscle and its tendon, but since their effective moment arms and muscle sizes are much smaller than the biceps, the hand forces, exerted by both wrist and fingers, are correspondingly smaller. The variation with wrist angle, of both flexor-extensor forces in the wrist and of prehensile forces in the hand, is of practical importance as well as theoretical interest. The prehensile force reaches a maximum at a wrist angle of about 145° (Fig. 5.10, bottom) which is approximately the angle at which the forces of wrist flexion and extension are maximum, which happens to be its normal angular position, and the position in which it assumes for strong grasps.

The fall-off in wrist strengths as the wrist flexes or extends away from the normal position correlates with the length-tension relationship for muscle as it undergoes stretch or slackening. The exception to this rule, seen in the augmented force of flexion at wrist angle 85°, apparently means that this degree of wrist extension does not stretch the flexor muscles beyond their force maximum.

With regard to UL prosthetic design, it would be possible to incorporate the natural dynamics of the wrist as shown in Fig. 5.10 by using position sensors and a processor, but this feature may or may not be advantageous for a prosthetic limb.

5.10 Hand Synergies

The bones, muscles, and tendons of the hand form a coordinated network whose dexterity is accomplished using "synergy." This feature, which essentially means simply, "working together," can be appreciated by examining Fig. 5.15, where it is seen that tendons and muscles of the extrinsic finger flexors and extensors cross several joints. This means that a single tendon can generate torques at several joints simultaneously. Thus the natural linkages between the individual bones of the wrist and digits can accomplish coordinated motions with a single volition from the brain. Moreover, since many muscles themselves are linked together passively, they are naturally coordinated. Thus the anatomical connections present in the arm can produce complex motions with minimal overhead on the part of the motor control system. Synergy, therefore, is the cooperation of several muscles and joints by a single neural drive. Without this arrangement, movements would require more conscious control. In terms of motor control, this arrangement reduces the dimensionality of the control space.

Synergy is clearly evident during development, when the newborn mammal can grasp reflexively long before he/she can individuate the digits. As the CNS matures, it can "override" this synergy to begin performing more intricate tasks such as typing and piano playing (Fig. 5.16).

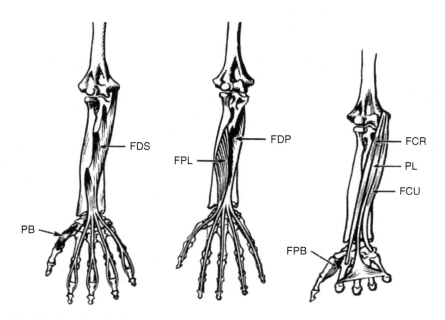

Fig. 5.15 Muscles of the hand

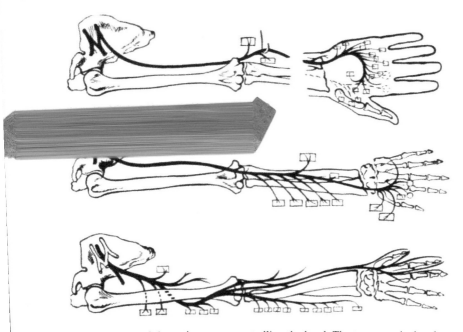

Fig. 5.16 Schematic of the major nerves controlling the hand. The top nerve is the ulnar nerve followed by the median nerve in the middle, and the radial nerve on the bottom. **CRAIG L TAYLOR, Ph.D.,1 AND ROBERT J. SCHWARZ, M.D., O&P Library.** The Anatomy and Mechanics of the Human Hand 3. 1955

Nerve	Muscles innervated
Ulnar	Flexor/extensor by intrinsic muscles of hand
Median	Flexors of wrist, thumb, fingers
Radial	Extensors of wrist, thumb, fingers

5.11 Exercises

1. Estimate the power and energy needed by a sound-limbed person to lift a 10 lb weight from neutral position (elbow fully extended) to 90°. Show and justify your assumptions. Hint: you will need to specify the contribution to the task by specific flexor muscles.
2. For the task in question 1, specify a specific motor, and its torque and power requirements, when used by a UL (BE) prosthesis.
3. For the task in question 2, add a closed-loop control system. Specify the components in a block diagram and show a simulation of the movement.

Chapter 6
The Human-Machine Interface

Goal: Learn options for controlling upper-limb prostheses.

6.1 Restoring UL Function

Requirements for restoring lost mobility to upper-limb (UL) amputees are: (1) a machine capable of mimicking UL motions, (2) user's ability to express a useful repertoire of volitions by natural or programmed signals, (3) a human-machine interface (HMI) that can register and execute those volitions.

The first requirement can be met by the many available multi-functional arms and hands, as exemplified by the NASA grip force-multiplying *Roboglove* (Fig. 6.1).

Requirement #2 can also be met since most amputees have some ability to express their motor volitions through nervous and/or muscular activity from either their residual limbs or at alternate sites on their body. The third requirement is not generally met, since reliably translating the user's signals into commands the HMI can understand has proven difficult. Thus the main challenge for UL functional restoration is the HMI (http://www.oandplibrary.org/alp/chap06-01.asp).

6.2 HMIs

Prostheses can be internally powered (IPP) directly by muscles or externally powered (EPP) by motors. The EPP motors are controlled by signals generated by residual muscles. These commercially available HMIs, and several HMIs still in the developmental stage, are categorized in Fig. 6.1.

As shown in Fig. 6.2, IPP mechanisms operate directly from the user's muscles. As shown, a user with above-elbow loss can make a variety of motions to operate the elbow. For example, a user with above-elbow loss (AE) uses bi-scapular abduc-

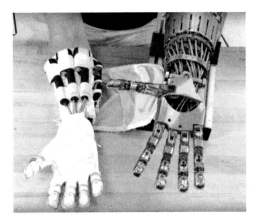

Fig. 6.1 NASA grip force-multiplying Roboglove. https://www.the-nref.org/content/nasa-grip-force-multiplying-roboglove

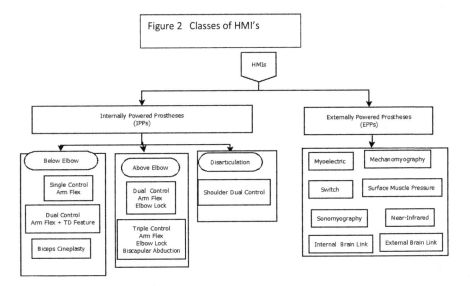

Fig. 6.2 Classes of HMIs

tion to operate a cable that opens the terminal device and can flex/extend the prosthetic elbow.

The HMI has three basic functions: (1) it physically attaches the prosthesis to the user, (2) it registers and transmits user's commands to the prosthetic hardware, and (3) feeds back force and position signals to the user. It is upon these channels of communication that the effective control of externally powered devices depends. Unfortunately, the biological channels are not well understood, and there are limited means with which to communicate with them. The basic strategy for improving the HMI consists of accurate sensing of user command signals, interpreting these signals, and producing the commanded volitions. A sampling of HMIs is described below.

Operation of internally powered arm prosthesis. Three different movements are accomplished by the body motions as shown.

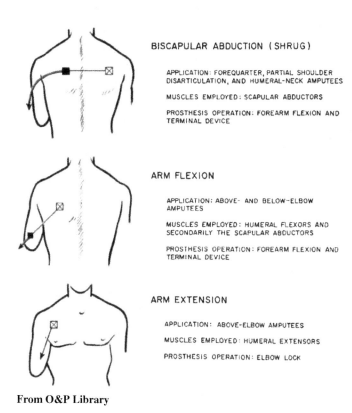

BISCAPULAR ABDUCTION (SHRUG)

APPLICATION: FOREQUARTER, PARTIAL SHOULDER DISARTICULATION, AND HUMERAL-NECK AMPUTEES

MUSCLES EMPLOYED: SCAPULAR ABDUCTORS

PROSTHESIS OPERATION: FOREARM FLEXION AND TERMINAL DEVICE

ARM FLEXION

APPLICATION: ABOVE- AND BELOW-ELBOW AMPUTEES

MUSCLES EMPLOYED: HUMERAL FLEXORS AND SECONDARILY THE SCAPULAR ABDUCTORS

PROSTHESIS OPERATION: FOREARM FLEXION AND TERMINAL DEVICE

ARM EXTENSION

APPLICATION: ABOVE-ELBOW AMPUTEES

MUSCLES EMPLOYED: HUMERAL EXTENSORS

PROSTHESIS OPERATION: ELBOW LOCK

From O&P Library

6.3 Example of an Advanced Internally Powered Prosthesis

One type of IPP is cineplasty: a procedure that surgically connects a muscle or tendon of the below-elbow residuum to the prosthetic hardware, which provides direct mechanical control of hardware, and more natural movements [1] (Fig. 6.3).

Fig. 6.3 Cineplasty. Force and excursion provided by the biceps muscle tunnel are harnessed by inserting into the tunnel a cylindrical pin of a nontoxic material and attaching a cable to each end of the pin. As in the other types of control systems, the Bowden cable principle is employed to maintain a constant effective distance between the source of energy and the mechanism to be operated, regardless of relative motions occurring between body segments. In order that conventional terminal devices may be employed, it is necessary to join the two cables before attachment to the mechanism. Several devices for making this coupling are available commercially

> The cineplasty tunnel in the biceps of the average male will provide sufficient force and excursion to operate modern terminal devices—an average maximum force of 50 lb. and 1½ in. of useful excursion. It is not unusual for some individuals to be able to build up the force available to a value in excess of 100 lb., but such a high force normally is not required (Weir, R.F., C.W. Heckathorne, *Journal of Rehabilitation Research and Development*, 2001. **38**(4): p. 357–363).

6.4 Externally Powered Prostheses (EPPs)

EPPs can be applied to all levels of amputation. Commercially available HMIs for EPPs use myoelectric (MYOE) sensing and decoding with various algorithms for extracting information from EMG signals. Prosthetic users can combine various signaling strategies, such as their MYOE signals and switches, i.e., by foot movement or cell phone signals, in order to optimize performance (Pylatiuk, Schulz et al. 2007, Resnik, Latlief et al. 2013). Many innovative strategies for HMI control are being developed, as discussed in this chapter.

6.5 Increasing Degrees of Freedom in the Upper Limb Prosthesis

The functions expected of an arm are many: typing, phoning, tooling, modeling, squeezing, self-care, et cetera. Each of these requires several degrees of freedom from the upper limb. For example, typing, even with one hand, requires at least two DoFs in the wrist, one in the elbow, and at least two in each finger. An HMI thus needs to control a minimum of eight simultaneous, or nearly simultaneous, DoFs for rudimentary typing, but available HMIs are far from capable of this. Available prostheses can deliver one or two DoFs and can produce a selection of different types of grasps, such as power, precision, and pinch, but are far from reproducing human dexterity.

6.5.1 Myoelectrical Signals as HMI Input

Ideally, the information encoded in the MYOE signals embodies the details of the user's volition and can be used to produce the volition by mechanical means. In other words, the volitional "encode" embodies the desired sequence of neuromotor commands. Each UL movement in space requires a specific set of muscular actions that activate and produce force, and another set of muscles that relax. Producing this motion artificially requires sensors that interpret the command and actuators that execute it. Extracting this information is the job of the amplifier and logic blocks shown in Figs. 6.4 and 6.5. MYOE signals can be recorded from one or more muscles that are interpreted by the signal processor. As shown in Fig. 6.4, an amplifier first conditions the MYOE signal by a process of filtering and rectification. Typically, MYOE signals are sampled at 100 Hz or more, and then band-pass filtered between 7.5 and 15 Hz and then rectified to yield the full power of the EMG signal.

In the simplest, and most common HMIs, the processed signals are compared against a pre-set value, such as a threshold amplitude in the simplest case, to establish a binary ON signal for motor control activation. The simplest and most common of these are simple thresholding. If, after filtering and conditioning, the signal meets the pre-set threshold, reliability of this paradigm is quite high, but better reliability can be achieved by combining signals from separate muscles, such as an agonist and antagonist pair, to form a vector, or a simple ratio, which is thresholded. This produces a more robust and detailed command. These strategies yield one DoF; however, greater dexterity, with multiple DoFs, can be achieved through higher-order signal processing. Schemes described in Sect. 6.9 can achieve even more specificity, and more dexterity from EMG signals, by extracting analog information from EMG signals that can modulate features such as force or velocity. Several laboratories, including the authors', have developed signal processing strategies in laboratory settings that have achieved better control. For example, by using just the early portion of the EMG signal rather than the entire duration of the signal,

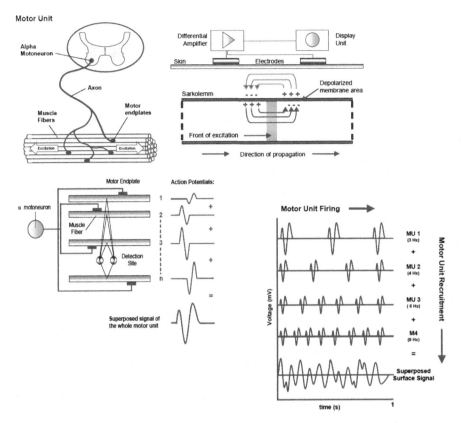

Fig. 6.4 Schematic example of recording myoelectric signals from four muscle fibers (left panel) and summing them (right panel). Peter Konrad: The ABC of EMG

better control can be achieved [2] (Natarajan, Wininger et al. 2011, Farina, Jiang et al. 2014) (Fig. 6.5).

As noted, surface EMG represents muscle electrical activity, which is used as an approximate measure of muscle force. Higher fidelity can be obtained from internally implanted EMG electrodes, whose output can be transmitted wirelessly to the motor controllers. This technique holds promise for better prosthetic control, since it eliminates problems with the skin-electrode interface and is more accurate; however the surgical procedure has drawbacks for many users.

Ultimately, the repertoire and quality of potential volitional movements is determined by the skill and patience of the user in producing correct muscle contractions and the abilities of the HMI to decipher each and accurately match it with the correct motion.

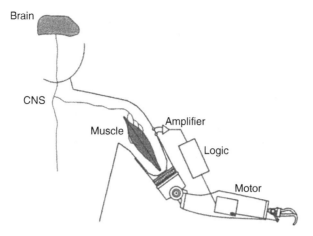

Fig. 6.5 Typical myoelectric HMI schematic. Downloaded from Digital Engineering Library @ McGraw-Hill (www.digitalengineeringlibrary.com) © 2004 The McGraw-Hill Companies. All rights reserved. Any use is subject to the Terms of Use as given at the website

6.6 Characteristics of MYOE

Electromyography registers muscle function through analysis of the electrical signals emanating during muscular contractions. These may be voluntary or involuntary muscle contraction. Myoelectric (MYOE) signals are widely used for UL prosthetic control; however they are often not reliable. Firstly, while MYOE signals are registered easily from sound-limbed persons, their registration from amputees is difficult, since amputees' residual muscles are generally damaged and fatty, yielding very low signal-to-noise ratio. The EMG signal is the algebraic sum of the motor unit action potentials within the pick-up area of the electrode being used. The pick-up (receiving) area of an electrode will almost always include more than one motor unit because muscle fibers of different motor units are intermingled throughout the entire muscle.

Most of the MYOE prostheses utilize a two-site agonist-antagonist control method. This is essentially an off-on paradigm, wherein the user activates a particular muscle, while trying to relax the antagonist, so the dual signal can be detected. When the signal exceeds a set threshold, the controller turns the motor on. This mode works adequately for many amputees, but it requires a degree of concentration from the user, since the required movement is far from natural. While the EMG is the established mode of controlling motorized prostheses, it is often unreliable and inappropriate for some users.

There are three approaches to registering EMG signals from a muscle: (1) surface electrodes, as depicted above, (2) percutaneous wires inserted by means of hypodermic needles, and (3) surgically implanted radio-transmitting devices. In the percutaneous method, wires protrude through the surface of the skin from inside the muscle. These provide a high resolution electrical record of the EMG, but are somewhat vulnerable to damage, and the wires must be flexible and provoke no ill effects on the tissue, such as infection or inflammation. Newer electrode materials and design may make this mode more practical.

6.7 Radio-Transmitting Electrodes Implants

Radio-transmission of EMG signals solves the percutaneous wire problem, but signal quality may not be as reliable. Electrodes, whether external or internal, are generally considered the weakest link in signal bio-transmission. Research is needed to find ways to prevent metal fatigue and to discover contact materials which produce no body-tissue reaction, and which do not corrode or weaken.

A telemetered HMI is shown in Fig. 6.6. Implanted electrodes are much more sensitive than surface electrodes and offer better resolution. The muscle signals are recorded with an external coil on the residuum and are wirelessly transmitted to motors. The IMES system offers improved limb control and has been tested in animals. Human testing is underway.

A variant of the IMES is a "myokinetic" hand controller, which records residual muscle activity with implanted magnets, whose movements are recorded with external magnetic field sensors, surrounding the residuum, as shown in Fig. 6.7.

6.8 Targeted Muscle Reinnervation

Patients whose residual muscles are too damaged may have a surgical option, whereby their hand motor nerves are rerouted to other sites on the body. This HMI strategy uses implanted electrodes in a procedure called targeted muscle reinnerva-

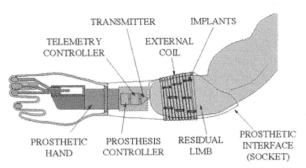

Schematic of planned Implanted Myoelectric Sensors (IMES) system.

Fig. 6.6 IMES. IEEE Trans Biomed Eng. 2009 January; 56(1): 159–171. doi:10.1109/TBME.2008.2005942, Implantable Myoelectric Sensors (IMESs) for Intramuscular Electromyogram Recording Richard F. ff. Weir et al

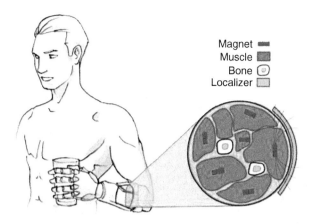

Fig. 6.7 Embedded Myokinetic Prosthetic Hand Controller. F. C. Valerio Ianniciello, M. Gherardini, and C. Cipriani *Sensors* 2019, 19, 3137; doi:10.3390/s19143137. The Biorobotics Institute, Scuola Superiore Sant'Anna, 56127 Pisa, Italy

tion (TMR). The procedure surgically redirects arm nerves in the residuum to sites at the *pectorals*, where the nerves re-grow and reinnervate (Fig. 6.8). TMR is relevant to high-level injuries, such as shoulder disarticulation, but its requirement for surgery and a lengthy (~2 years) recovery, followed by placement of a large number of EMG electrodes on the chest, makes it a daunting ordeal. The chest is used for this procedure because it presents a large, flat area for recording electrodes. To explore better options for exoskeletal control, a scientific panel, the Peripheral Nervous System-Machine Interface (PNS-MI) group, is addressing the problem.

As seen, each of the five gesture produces a highly distinct map of EMG amplitudes, registered by a dense array of electrodes on the chest. The color maps are coded for amplitude, with the darker red indicating locations of high amplitude, and the lighter colors indicating smaller amplitudes. Each gesture request can thus be distinguished by the electrode patterns, which are translated into the desired prosthetic motion. Depiction of the mapping of the arm nerves to sites on the pectorals is shown in Fig. 6.9.

6.9 Vectorializing MYOE Signals

A promising improvement in MYOE control uses a vectorial approach to the EMG signals. As shown in Fig. 6.10, multiple EMG simultaneous signals from several sites on the arm are recorded while the user executes several gestures (Jacob L. Segil, Richard F. *ff.* Weir) [3]. This approach records from 12 separate sites around the circumference of a limb, in this case the arm, and a resultant EMG vector is calculated from the combination of each signal, for each. Each gesture produces a different resultant vector that can code for specific motions. The method involves placing

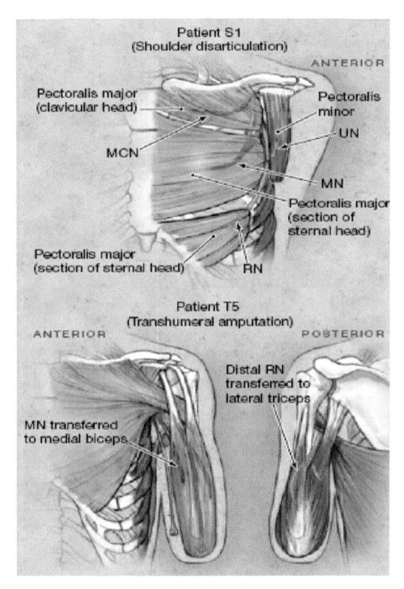

Fig. 6.8 Sketches of pectorals, and arm nerves, that can be rerouted for better signal transmission. Kuiken, T

12 electrodes around the BE residuum, so EMG rms magnitude of each site represents a component of a 12-dimensional vector in the xy plane as shown. In this way, for any given combination of EMG signals, a resultant vector, R, can be computed. Thus, the number of DoFs available to the prosthesis is limited only by the number of distinguishable vectors the user can reliably produce by particular muscular

6.9 Vectorializing MYOE Signals

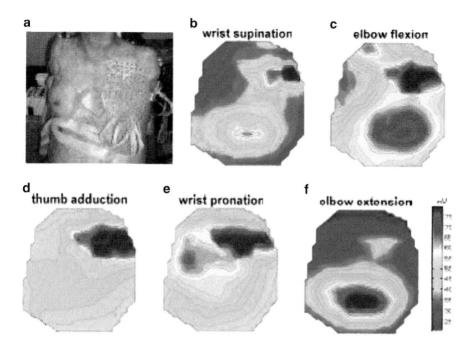

Fig. 6.9 (a) Electrodes on pectoral muscles which are innervated by nerves from the arm. (**b–f**) Results from TMR. The contour plots show the spatial distribution of EMG amplitudes for the various movement volitions. Kuiken, T

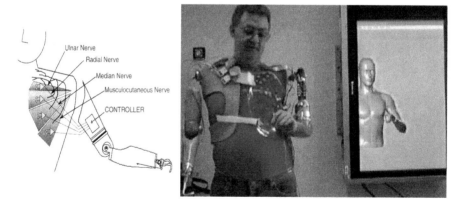

Fig. 6.10 Schema of a TMR Controller and a user manipulating it

actions. Able-body subjects could reliably produce six DoFs, consisting of the different hand grasps shown in Fig. 6.9d.

The summation of all vectors produces a resultant vector (R):

$$\vec{R}(t) = \begin{bmatrix} R_x(t) \\ R_y(t) \end{bmatrix} = \sum_{i=1}^{N} \begin{bmatrix} \text{RMS}_i(t)\cos\theta_i \\ \text{RMS}_i(t)\sin\theta_i \end{bmatrix}$$

By vectorializing several EMG signals, the resultant, R, embodies much more information compared with a single EMG amplitude. As shown, six different grasps can be coded. In each of the three examples of high resolution MYOE recordings, the process involved the user training the controller to recognize and decode the patterns (Fig. 6.11).

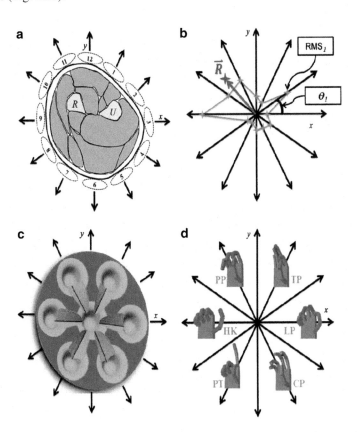

Fig. 6.11 Vectorial decomposition of EMG signals. (**a**) Untargeted electrode array (1–12) is arranged about cross section of the forearm. R radius bone, U ulnar bone. (**b**) Vector summation map depicts root mean square (RMS) electromyography activity as measured by electrode array. Vector summation algorithm calculates resultant vector ( ). θ = joint angle array. (**c**) Example potential field design where red/green areas distinguish areas of zero/negative potential, respectively. This potential field design was used in Experiment B. (**d**) Exemplary postural map with seven postures arranged in symmetric distribution about postural control domain (hand flat posture not shown at origin). This postural map design was used in Experiment B. *CP* cylindrical prehension, *HK* hook, *LP* lateral prehension, *PP* palmar prehension, *PT* pointer, *TP* tip prehension (Jacob L. Segil, Richard F. ff. Weir) [3]

Voluntary movements begin in the motor cortex and travel through central processing circuits, nerves, and muscle and are expressed as kinetic activity in the residuum. Thus mechanical activity on the residuum surface represents volition after degradation through the entire system. Volition can thus be characterized by a degradation function, and its inverse, a restoration function, can discriminate specific movements.

6.10 Direct Brain Control

Direct brain-machine interfaces (BMIs) have been implemented on primates and several amputee volunteers using implanted cortical electrodes [4]. The technique involves implanting electrodes into regions of the cortex that may control specific limbs, as depicted in the homunculus in Figure and the highlighted region shown in Fig. 6.12. Experiments with primates have demonstrated their ability to learn control of a robotic arm via brain signals. Thus far, the practicality of BMIs is unproven and is difficult to implement.

The BMI consists of an array of mini-electrodes, as depicted in Fig. 6.12, that is plunged into an area of the cerebral cortex that is likely to emit motor control signals for the arm. The electrodes may record from a relatively small area, i.e., 1 cm^2, with an array size of 128 electrodes or more. The recordings are decoded, using an algorithm, such as the vectorial decomposition shown in Fig. 6.12. Most of the trials

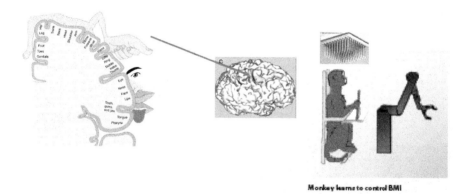

Fig. 6.12 Schema of direct brain recording. Homunculus at left is used to locate electrodes near the arm region. Sketch at right shows the test subject (monkey) controlling a robotic arm. Electrode array is shown (magnified) at top. This is plunged into the cerebral cortex. https://r.search.yahoo.com/_ylt=AwrEzNxG8MpdllkAkFyjzbkF;_ylu=X3oDMTBtdXBkbHJyBHNlYwNmcC1hdHRyaWIEc2xrA3J1cmw-/RV=2/RE=1573609670/RO=11/RU=https%3a%2f%2fen.wikipedia.org%2fwiki%2fCortical_homunculus/RK=2/RS=RlGA2V9KIAl4A7Y0m8jnAsCnAwk-. Joseph T. Belter, MS, BS; Jacob L. Segil; Aaron M. Dollar, PhD, SM, BS; Richard F. Weir JRRD. Volume 50, 2013

with BMI have been with monkeys, who were trained to control a robotic arm to reach for a reward.

The BMI interface is difficult to implement and has not yet yielded much functionality. For example, it was recently noted that a BMI user "was able to obtain three degrees of freedom control in less than a month" [4]; while this feat is a breakthrough, especially for high-level injuries, similar results can now be attained intuitively by directly interfacing with peripheral nerves (PNS-MI) [5, 6]. Further details on the BMI interface are presented in Chap. 10.

6.11 Alternative HMIs

The HMI must reliably extract and execute the user's volition from the user's signals. In the case of MYOE signals, there are two major challenges : (1) the MYO signal is composed of electrical noise which is difficult to record reliably, and (2) the residuum producing the signal is often damaged and scarred, and thus not capable of generating clear signals. As seen in the MRI and photographic images of a residuum that resulted from electrical burns, the muscular tissue is damaged and largely replaced by scarred and fatty tissue (Fig. 6.13).

Due to these major limitations of the MYOE, alternative HMI method has been developed by recording the mechanical, instead of the electrical signals from the muscle. Abboudi et al. [7] showed that amputees could achieve better control of their UL prosthesis using a HMI based on mechanical signals from their residua. The method uses a sensory interface, based on force myography (FMG), which registers the dynamic pressure and shape changes within the residuum associated with specific movement volitions. These three-dimensional mechanical dynamics can be measured with myo-pneumatic (M-P) or force-myographic (FMG) sensors within the socket [8]. A fine degree of residual limb muscle control was also demonstrated by amputees in a study using Hall effect magnetic movement sensors [9] (Fig. 6.13).

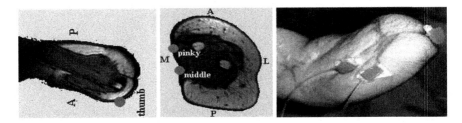

Fig. 6.13 MRI and photographic pictures of a residuum [7]. Left panel shows a MRI image, sagittal, of the residual limb; middle panel shows a cross section MRI; right panel is a photograph of the residual limb, with sensors. Images of the residuum of an amputee subject who lost the limb in an electrical accident

The *Dextra Hand* was introduced in 1999, offering multifunctional control of a prosthetic hand, a new controller based on the principle that the mechanical actions of muscles, both agonist and antagonist, as measured by surface muscle pressure [10]. This mode is essentially force myography (FMG), which, unlike the EMG that registers the electrical signals, expresses the entire volitional signal, including trajectory and force, in a highly processed and invariant manner. FMG codes well for the motion and force of several DOFs, in users who retain some residual arm muscles.

Compared with the EMG, the FMG is a more robust and accessible HMI, since it decodes the mechanical dynamics of residual muscle, both agonist and antagonist, which encode volition more directly than does the EMG [7]. Surface muscle pressure (SMP), as registered by FMG, expresses the entire volitional signal, including trajectory and force, in a highly processed and invariant manner. For users who retain some residual arm muscles, SMP accurately codes for the motion of several DOFs, as well as force; its advantages include (1) inherently more reliable and stable signals, (2) resolution that is not dependent on precision of sensor placement, allowing for convenient donning and doffing of the socket, (3) insensitivity to sweating, and (4) biomimetic, intuitive control. The SMP controller, *Dextra*, was used by a person with trans-radial amputation playing a short piano piece in a BBC special and other media after only a few minutes training. A recent review stated that "beyond providing dexterity, the Tendon-Activated Pneumatic Hand [AKA *Dextra*] controller may facilitate the transition to more complete hand restorations over the future". Several other HMI options are available to try, including ultrasound, optics, and NIRS.

6.12 Comparing FMG with EMG

Force myography (FMG) records the surface muscle pressure (SMP) exerted by dynamic contractions. To verify their validity as a representation of muscle activity, FMG signals were directly compared with EMG as shown in Fig. 6.14. A subject was instrumented with both types of sensors on a leg and was directed to flex the thigh. Simultaneous SMP and SEMG records from isometric contractions are shown during knee extension, comparing Butterworth low-pass filter (bottom panel) and 7-point moving average filter (top traces) and raw signals (lower traces). Panel (A) SEMG records from 6 s of *quadriceps* isometric knee extension. (C) Superimposed and expanded SEMG and SMP signals from panels (A) and (B) during knee extension.

As seen in Fig. 6.14, FMG accurately records the timing and magnitude of volitions, because it captures the mechanical pressure from the muscle and tendon tension. Note that the FMG signal is less noisy compared with EMG, making it a better measure of volition.

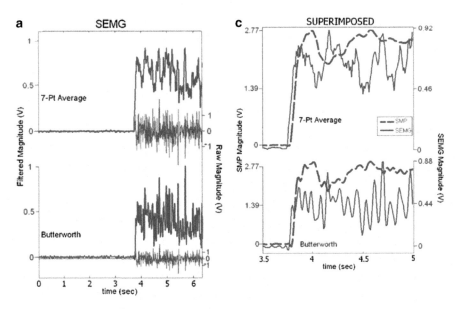

Fig. 6.14 Comparison of SMP and surface EMG signals. This is from Craelius Lab

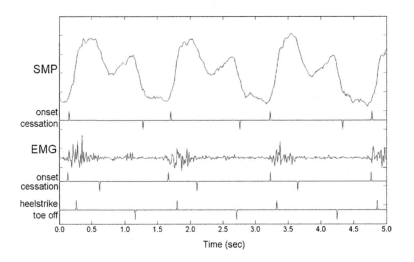

Fig. 6.15 Comparing SMP signals with EMG

Another comparative test of volition sensing is shown in Fig. 6.15. A subject was instrumented with both types of sensors on a leg and was directed to flex the thigh repetitively. Simultaneous SMP and SEMG records from isometric contractions are shown during knee extension; Panel (A) SEMG records from 5 s of *quadriceps*

6.12 Comparing FMG with EMG

isometric knee extension. (C) Superimposed and expanded EMG and SMP signals from panels (A) and (B) during knee extension.

The data show that FMG (red signal) provides more accurate records of the timing and magnitude of volitions, because it captures the mechanical pressure from the muscle and tendon tension, which is a much less noisy signal. Note that the signal-to-noise ratio of the SMP is much higher than that of the EMG. The signal is less noisy compared with EMG (blue signal), making it a better measure of volition.

FMG has been adopted for many other HMI applications involving prosthetic control and muscle recording. FMG can use sensors made from foam, force-sensitive resistors, or from optical fibers (Fig. 6.16).

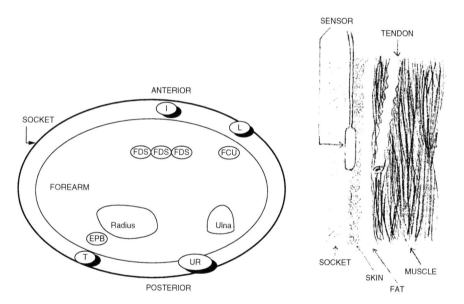

Fig. 6.16a Sketch of sensor arrangement on the residuum. Pneumatic FMG sensors are fabricated from porous polyurethane foam, vacuum formed within a polyethylene bag and attached to a flexible tube. Alternate sensors can be force-sensitive resistors [7]

Response Matrix
T, I, L represent sensor <u>outputs</u> from Thumb (T), Index (I) and Little (t, i, l represent requests from thumb, index and little, respectively.

Request	T	I	L
t	tT	tI	tL
i	iT	iI	iL
l	lT	lI	lL

Fig. 6.16b Confusion Matrix of FMG Volitions

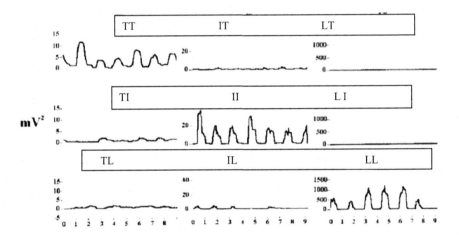

Fig. 6.16c Signal response matrix obtained from an amputee over a 9 s period. Traces represent squared signals derived from SMP sensors. Amplitude scale for each sensor (along the column) is constant, but the scales were adjusted for each sensor. Note that the signals along the diagonal correspond to the correct sensor being activated by the movement. Minimal signals in the off-diagonals indicate the high specificity of the controller

6.13 Developing a FMG-Based HMI

Sensors were arrayed around the residual forearm as shown below. To optimize signals, the muscles were palpated during volitions to identify sites of high activity. The test involved the subject trying to perform taps of three different fingers of the prosthetic hand.

6.13 Developing a FMG-Based HMI

Testing of FMG as an HMI for upper-limb prosthetic control.

At right, a cross section of sensor placement. Targeted locations for sensors are based on desired motion. Thumb motion (T) is sensed posteriorly via the extensor pollicis brevis (EPB). Index (I) and little finger (L) motion are sensed anteriorly via the flexor digitorum superficialis (FDS) and the flexor carpi ulnaris (FCU), respectively. Ulna/radius rotation (UR) is sensed at the intersection of the two bones. Suitable sensor locations were circumscribed by palpating the forearm during volitional finger motions. These preliminary sites corresponded to specific tendon movements as illustrated in the forearm cross section above.

RESPONSE MATRIX. T, I, L REPRESENT SENSOR OUTPUT FROM THUMB, INDEX, AND LITTLE DIGITS, RESPECTIVELY;

t, i,l REPRESENT REQUEST FROM THUMB, INDEX, AND LITTLE DIGITS, RESPECTIVELY.

Signal response matrix obtained from an amputee over a 9 s period. Traces represent squared signals derived from TAP sensors. Amplitude scale for each sensor (along the column) is constant, but the scales were adjusted for each sensor. Note that the signals along the diagonal correspond to the correct sensor being activated by the movement. Minimal signals in the off-diagonals indicate the high specificity of the controller.

Current versions of the FMG-HMI use silicone sleeves loaded densely with sensors, in the instance, force-sensitive resistors (FSRs). The sleeve is custom fit for the residuum. The circumferential array of up to 36 sensors provides a 14-dimensional "force map" of the entire residuum (Fig. 6.17), which yields many degrees of freedom for operating a prosthesis.

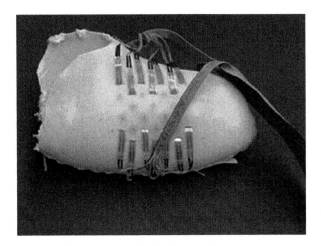

Fig. 6.17 FMG Sensor Sleeve. Fourteen sensors are located within the circumference

6.13.1 Silicon Sleeve with Integrated FSR Sensors

The FMG provides a rich map of the residual surface, yielding a high resolution volitional signal that includes details of the volition. For users who retain some residual arm muscles, FMG accurately codes for the motion and forces of several DOFs. A recent review stated that "We found that FMG yields in general a better performance than sEMG"; Nowak, Eiband, Ruiz Ram and Castellini; Journal of Neural Engineering, 2020. Several other HMI options may also be advantageous and remain to be implemented, including ultrasound, optics, and near-infrared signals (NIRS).

With sufficient density of sensors on the residuum, muscle activity maps can be calculated, as shown at left. Here the amputee subject was asked to make volitions for thumb, middle, and pinky, while his myokinetic forces were recorded and mapped in gray scales. The scale ranged from black to white, denoting forces from low to high. Each map was vectorialized, as depicted below, and each volition was decoded using a filter vector, into a control vector, that executed the movement on the robotic hand (Fig. 6.18).

FMG signals can be obtained via any force sensor.

For example, optical fiber transducers measure force based on principles of microbending, and they seem to provide similar information as FMG signals. The microbending transducer was applied to gesture recognition for the real-time control of a virtual prosthetic hand. The sensor successfully classified four hand postures, similarly to the results shown above for FSR and foam sensors. Thus FMG can provide a simple low-cost and reliable approach for controlling multi-grasp bionic hands (Fig. 6.19).

6.13 Developing a FMG-Based HMI

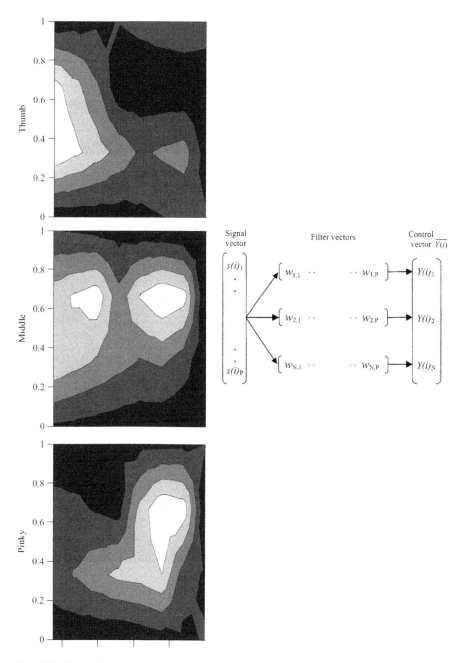

Fig. 6.18 Each of the images represents a vector containing a "residual kinetic image" of residual limb movements for subject A during requested finger tapping. Each requested movement is identified by a unique image of pressure energies. Maximum pressures were approximately 3 kPa with white being the greatest and black being no change. David J. Curcie, James A. Flint, and William Craelius *IEEE TRANSACTIONS ON NEURAL SYSTEMS AND REHABILITATION ENGINEERING, VOL. 9, NO. 1, MARCH 2001*

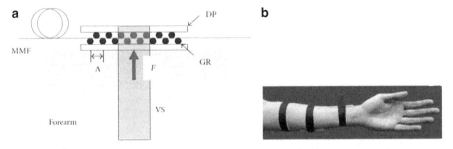

Fig. 6.19 (**a**) Schematic of the optical fiber FMG transducer. *MMF* multimode fiber, *DP* deformer plate, *GR* graphite rods, *VS* velcro strap, *Λ* microbending periodicity, *F* applied force. (**b**) Image of the transducers attached to user forearm. Eric Fujiwara and Carlos Kenichi Suzuki

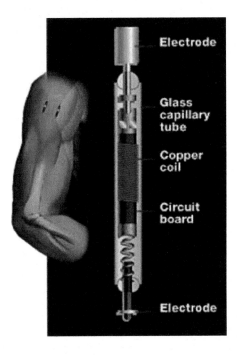

Fig. 6.20 Multimodal injectable sensors for neural prosthetic proprioception. Tan, W (Tan, W); Sachs, N (Sachs, N); Guo, R (Guo, R); Zou, Q (Zou, Q); Singh, J (Singh, J); Loeb, GE (Loeb, GE), 2005 First International Conference on Neural Interface and Control Proceedings. Edited by:He, JP; Gao, SK; Lin, JR Pages: 172–176 Published: 2005

Another sensing modality is the Bion (Fig. 6.20), invented by Gerald Loeb. This device is the size of a grain of rice and can be injected into a muscle for registration of information on position and muscle activity. It communicates the information to a small external processor which can be useful for controlling a prosthesis.

6.14 Biomimetic Control

The closest alternative to biological control by nerves is biomimetic control, using a high resolution HMI, similar to that described above. The HMI needs to exploit the presence of functional actuators within the residual limb that once controlled the wrist, hand, and fingers. These include the extrinsic muscles and tendons that once controlled the flexion of the metacarpal-phalangeal joints. While myoelectric recording from the individual finger-associated muscles, i.e., separate branches of the flexor digitorum, is not possible with surface electrodes, sensing the mechanical activity of the muscles and their associated tendons is possible. These recordings potentially contain more coherent information about volitions, since it is individual tendons that directly move the joints.

A biomimetic HMI described here is based on FMG, with pneumatic or force-sensitive resistors, to transduce tendon motions into control signals for the upper-limb actuators that remain intact in the residual limb. These include the extrinsic muscles and tendons controlling flexion of the metacarpal-phalangeal joints. The prosthesis can operate in two modes, binary, for single digit tapping, or proportional, for smooth, controlled motions (Fig. 6.21).

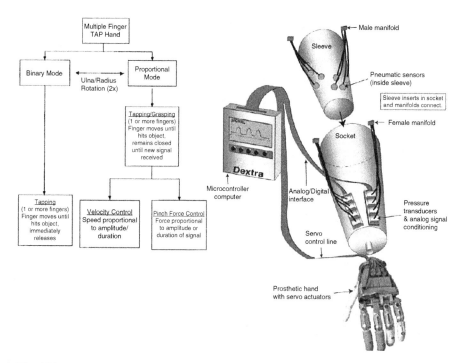

Fig. 6.21 DextraHand. Craelius, *Science*, 2002. www.dextrahand.org

> "The controller described here uses FMG to register muscle and tendons in the residual forearm and use them to control the tapping of multiple fingers of a mechanical hand. With further development this concept was developed into the dextra hand, allowing an amputee to play the piano two-handed, as shown in a BBC broadcast. FMG provides a more reliable control signal than does MYOE."

> High-density force myography: A possible alternative for upper-limb prosthetic control; Radmand, Ashkan; Scheme, Erik; Englehart, Kevin; JOURNAL OF REHABILITATION RESEARCH AND Development Volume: 53 Issue: 4 Pages: 443–456 Published: 2016.

While HMIs have advanced over the past three decades, these have been mainly developments of smaller components and faster processors. Improving communication between prosthesis and user remains a challenge. Restoring intuitive control of limbs requires not only fine motor control, but also sensory feedback to the user. For adequate control, the HMI requires "closed loop" that returns afferent signals about touch, force, texture, and motion, derived from the muscles and their tendons. Myoelectric control is a beginning, but it taps the middle, not the beginning, of the efferent loop. Current HMIs do not know the output of the system; this can only be inferred rather than known. A more intuitive control system is the next hurdle for prosthetics designers. Many sensory inputs could be incorporated to monitor ongoing position, force, and velocity, to better mimic the natural neuromotor system.

6.15 Volitional Signal Processing

Common motions are the result of direct volitional signals from central command (motor cortex) combined with pre-patterned movements which are coordinated at levels below the conscious nervous system. The prosthetic HMI must convert these low-level and generally noisy control signals that are generated by the user into coherent, accurate, high-level inputs for the prosthetic controller. In order to replicate user motion volitions with high fidelity, the HMI should embody systems that mimic the human control system, primarily its feedback mechanisms, as are detailed in Chap. 7. In other words, the HMI must operate in a "closed loop" manner, whereby sensory information regarding position and velocity is continually used to help direct the movement. With a closed loop system, the kinematics would be under the control of the processor, with the goal of achieving finer movement control. With the ideal control system, the user provides the command, and the controller adjusts the motor outputs to smoothly and accurately achieve the target. Thus the

6.15 Volitional Signal Processing

controller replaces what the lower motor neuron system does for the human being. Such technology is rapidly advancing the performance of orthotics and prosthetics systems. The main challenge to better control is the information-channel capacity, i.e., bandwidth, of the HMIs.

It should be noted that most of the elements of man-machine systems apply to both prosthetic and orthotic devices, but, when output systems are considered, it is necessary to discuss orthoses and prostheses independently, except for certain communication devices which apply to both. For example, much effort has been devoted to modifying telephone, recorder, typewriter, radio, and television equipment for easier use by handicapped persons. Touch dialing, alone, is an important asset, and many other activities of daily living can be aided by man-machine systems.

An important function of a hand is controlling strength of grasp, ranging from a very soft grasp for delicate objects and a hard grasp for heavy objects. To measure the ability of FMG signals to encode grip force, subjects were tested using the setup as shown in Figs. 6.22 and 6.23.

$G(s)$ and $Y(s)$ denote the convolution of the Laplace-transformed grip force dynamometer sensor outputs onto Fourier-domain representation of linearization equation (Eq. 3). These signals were transformed by Laplacian operator (denoted by L) and converted back into time domain by inverse Laplace (L^{-1}). Low Pass = second-order Butterworth low-pass filter with 4 Hz cutoff, Rect = rectified (Fig. 6.24).

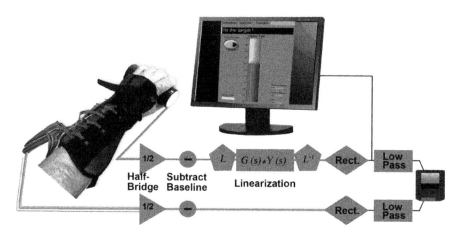

Fig. 6.22 Setup and processing for testing the ability of FMG to predict grip force. Subjects wore a FMG sleeve instrumented with 16 force-sensitive resistors, while actual grip force magnitudes were displayed on the screen as they squeezed a grip force dynamometer. $G(s)$ and $Y(s)$ denote the convolution of the Laplace-transformed sensor outputs onto Fourier-domain representation of a linearization equation. These signals were transformed by Laplacian operator (denoted by L) and converted back into time domain by inverse Laplace (L^{-1}). Low Pass = second order Butterworth low-pass filter with 4 Hz cutoff, Rect = rectified. Michael Wininger, Nam-Hun Kim, William Craelius, Journal of Rehabilitation Research & Development DOI: 10.1682/JRRD.2007.11.0187

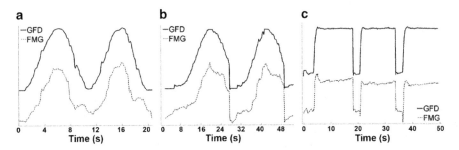

Fig. 6.23 Sample records of grip force dynamometer (GFD) and force myography (FMG) from different protocols. Graphs (**a**) and (**b**) represent subject applying a semi-sinusoidal grip, and (**c**) represents isometric grasps sustained at 50% maximum voluntary contraction for 10 s (three cycles). Journal of Rehabilitation Research & Development, DOI: 10.1682/JRRD.2007.11.0187, Michael Wininger, BS; Nam-Hun Kim, PhD; William Craelius, PhD

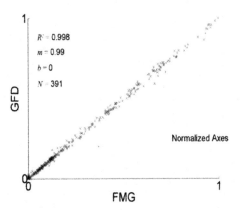

Fig. 6.24 Instrumentation and signal processing flowchart. Plot of linear regression to temporal landmarks of force myography (FMG) detection of forearm muscle pressure vs. grip force dynamometer (GFD) detection of grasp force. Across all protocols, subjects exhibited linear slope of $m = 0.99$ and intercept of $b < 0.001$, fitting 391 grasps. Scale normalized for clarity

6.16 Prosthetic Developments

The historical development of artificial limbs is a fascinating study of opportunistic events, as discussed in Chap. 1. Although seemingly simple and perhaps crude, the cable-controlled, rubber-band hooks commonly used by below-elbow amputees for over a century are, in fact, quite sophisticated, and many amputees have developed remarkable dexterity with them. Probably many years will elapse before the users of muscle or brain-controlled, electrically powered hands achieve the same level of reliability and dexterity now found in thousands of skilled hook users around the world.

The HMI problem is severe for users with above-elbow or bilateral amputations. For these clients, prosthetic functional restoration is marginal, such that the small increase in function provided by the prosthesis is often deemed too small to make it worth his while to learn to use it. The most successful systems to date are powered by compressed gas or electric motors. Each clinical application represents a major engineering achievement, and each one is usually somewhat different from all others. This is the real limitation in the development of sophisticated upper-extremity systems, for the problem of fitting and the nature of disability are so different among the relatively limited numbers of amputee users and congenitally deformed children that the sophisticated engineering required is often economically unjustified. However, the obvious challenge presented by the creation of an artificial human limb continues to fire the imagination of engineers throughout the world, and one may expect continued progress.

The case for the lower-extremity prostheses is somewhat different, because a man cannot walk with just one leg. Much effort has been devoted to developing lower-extremity prostheses for both above-knee and below-knee amputees. A successful prosthetic application requires close collaboration between the orthopedic surgeon and the prosthetist. Thoughtful planning concerning the nature of the stump to be produced can make a great difference in the effectiveness of the final prosthesis. Walking is one of the most sophisticated patterned functions in man. Many muscles are interrelated in producing a gait of minimum energy expenditure. One area of intensive research has been the study of human gait in order to improve the design of lower-extremity prosthetic and orthotic devices.

Lower-extremity prostheses are surprisingly complicated. Instead, one must build nonlinear damping devices into a lower-extremity prosthesis to control the swing phase so that it will approximate that of a normal human being. In the simplest versions, disks of leather are used to provide this friction. Recently, nonlinear and hydraulic devices have been built into artificial limbs. These hydraulic devices still suffer occasionally from seal and other failures, but they have been successfully used by amputees under a Veterans Administration evaluation program.

The problem of socket design and fitting is still under investigation, for one must transfer considerable forces to the limb, both in direct compression and in torsion. Sockets providing total surface contact, air cushions, "breathing effect," and special types of support have been developed. For a number of years, researchers have attempted to measure the pressure distributions occurring under dynamic conditions within lower-extremity sockets. In general, these attempts have not been successful, and this remains a challenging area for future research. Such pressure distribution data are urgently needed for the intelligent design of lower-extremity prostheses and, in some cases, for upper-extremity devices.

6.17 Sensory Receptors and Local Feedback

A user controlling either the most simple or the most complex assistive device requires feedback information to achieve her goal. In normal human motor activity, feedback comes via sight, sound, touch (pressure), and proprioceptive senses. These normal feedback channels are always impaired to some degree in handicapped persons and may be altogether missing. The visual path is still the most important for control in most orthotics and prosthetics systems, but now technology can relieve the user of the need to keep his eyes consciously fastened on each part of an output task. Additionally, users can hear the sounds of the prosthetic motors as feedback cues for feedback control, which may or may not require conscious reception by the subject. Many amputees learn to interpret reflected forces and motions through Bowden cables and other body-powered components.

Feedback control of orthotics and prosthetics systems is limited by the inability to provide effective artificial sensory feedback and will constitute a major barrier to overall system effectiveness for some years to come. It seems clear that a maximum research effort should be made to develop effective pseudosensory feedback signals, not only for orthotics and prosthetics systems but also for sensory aids to the blind and deaf—areas which are, of course, closely related.

Restoring some degree of sensory feedback to users can improve performance greatly. For example, as a substitute for touch sense, sounds can be used for indicating pressure magnitudes. Also, skin vibrations can be transmitted to indicate position. The phase and intensity of vibrations at two points on the surface of the skin can cue spatial position. Many other sonic strategies and sensory substitutions can help restore spatial and other manual sensations to users.

6.18 Adaptive Neuromotor Learning

One of the great virtues of biological systems is their ability to adapt, a prime example of which is the adaptability of our neuromotor system in controlling our joints according to our volitions. In each individual, this ability has been learned by our neuromotor system, during development from infancy to adulthood. The learning process involves specific neuronal connections from the brain to the spinal cord to our muscles. Natural dexterity is thus the product of an adaptive process that formed our control system during development.

While adaptation to our natural limbs occurs naturally over a long period, adaptation to a prosthesis is an unnatural process that needs to be quick. A procedure known as *implicit motor control training* (IMCT) has proven to be an efficient tool for helping amputees adapt to their new limb. The concept is based on the idea that the user should teach his/her device how to execute volitions, so that the prosthesis is the learner and the user is teacher, rather not the other way around. An example of IMCT involved UL amputee subjects training a computer to recognize their hand volitions registered by a sensor sleeve [12]. The subject practiced a standard, "9-hole peg" test in a virtual environment, as shown in Fig. 6.25a. Preliminary results with IMCT are positive, and have shown that amputees can readily teach their virtual hand to perform the manual task, with scores on par with control subjects. In another type of IMCT, the user controls and navigates a robotic arm, as shown in Fig. 6.25b (Dr. Artemiadis and colleagues [13]).

 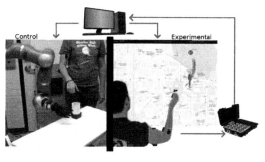

Fig. 6.25 Training upper-limb prostheses. At left, an amputee user practices the 9-hole peg Test, with visual feedback. Kuttiva, M., Flint J, Burdea, G, Craelius, W, *VIA: A virtual interface for the arm of upper-limb amputees.* 2nd International Workshop on Virtual Rehabilitation: IWVR, 2003. **2**: p. 119–126. At right, a sensory feedback system is used to help guide arm motion. Antuvan, A, Ison, CWM, Artemiadis, P, *Embedded human control of robots using myoelectric interfaces.* IEEE transactions on neural systems and rehabilitation engineering: a publication of the IEEE Engineering in Medicine and Biology Society, 2014. **22**(4): p. 820–827.

The difficulty for users controlling a device increases as the number of degrees of freedom (DoFs) it employs. Each DoF requires a specific, unambiguous signal from the user. As described in One may visualize a multi-axis orthosis controlled by EMG signals from six or more voluntarily excited muscles. An unanswered question remains as to how well the user can train the six or more muscles to perform the functions required, especially when the functions may be very different from those for which the muscle was naturally used. It is important to understand the difference between so-called naturally conditioned communication channels (NCCC) and operant-conditioned communication channels (OCCC). It appears intuitively correct to use the naturally conditioned channels wherever possible as signal sources for natural functions. The EMG-controlled artificial hands previously referred to use signals obtained from the prehensile extensors and flexors so that the amputee may open and close his artificial hand with the same muscles he would have used prior to the amputation. However, as degrees of freedom increase and the nature of the disability precludes naturally conditioned sources, one is forced to employ other muscles, such as the auriculares muscles behind the ears or the trapezius muscles in the shoulders, as signal sources.

It is clear that much research on these issues remains to be done. Age is certainly an important factor, for it is known that young children adapt very much more easily to orthotic and prosthetic devices than do older persons. Learning capability is closely related to the amount of information being received by the user through his feedback channels and to the amount of patterning and programming that can be done at the signal processing level. No doubt the future will bring clarification of these matters.

6.19 Cosmetic Restoration and Agency

While the HMI is generally a limiting factor for humans' interaction with machines, it is particularly troublesome for amputees, whose sensorimotor control system does not readily adapt to limb absence. Specifically, amputees often experience cognitive and sensory dissonance when perceiving an absent limb; these phenomena are expressed in the brain and are likely associated with phantom limb (PL) pain [14, 15]. The nature and possible neural origins of PL pain have been intensively researched [7] and are associated with brain regions that respond to incongruence between volition and perceived action [8, 9]. In addition to noxious pain, some amputees experience "referred phantom sensations," wherein stimulation of sensors on a prosthetic hand causes homologous sensations in the phantom hand [16]. The anterior insula in particular is a brain region that activates during motor-sensory conflicts, which are associated with error agency, i.e., a loss of body ownership or "selfness" [17–19]. Insula activity has also been associated with pain [20]. While error agency is reflected in the insula, a positive sense of agency correlates with activity in the parietal and prefrontal cortical areas as measured by a rise in oxyhemoglobin levels, in response to a real-time virtual tool experiment [5]. Further evidence of parietal cortex involvement with agency is provided by a study on amputees in which the size of their posterior parietal cortex correlated inversely with their amount of prosthetic usage and also with the intensity of PL pain they experienced [21]. These results underscore the role of agency in PL sensations and its localization to particular brain regions.

Strategies to mitigate PL pain by exploiting the brain's response to agency have included mirror imagery [4, 22–24] and sensory feedback from surrogate and virtual limbs [6, 11, 25–27]. A contrasting result was reported in a recent study wherein PL pain intensified when amputee subjects used their brain signals to control an advanced prosthetic hand; surprisingly, pain was reduced when subjects used a brain-machine interface to dissociate their use of a prosthetic hand from their PL [28]. Thus, manipulating perception can profoundly influence clients' neural activity, which may or may not be desirable.

Acknowledgments National Academy of Sciences (no copyright)
The O&P Library (no copyright)
James B. Reswick, Sc.D., Lojze Vodovnik, D.Sc., D. s. MCKENZIE, M.D. (published in O&P Library, no copyright)

References

1. Weir RF, Heckathorne CW, Childress DS. Cineplasty as a control input for externally powered prosthetic components. J Rehabil Res Dev. 2001;38(4):357–63.
2. Boostani R, Moradi MH. Evaluation of the forearm EMG signal features for the control of a prosthetic hand. Physiol Meas. 2003;24(2):309–19.

References

3. Segil JL, Weir RFF. Novel postural control algorithm for control of multifunctional myoelectric prosthetic hands. J Rehabil Res Dev. 2015;52(4):449–65.
4. Wosnitzka M, et al. Mirror therapy for the treatment of phantom limb pain after bilateral thigh amputation. A case report. Schmerz. 2014;28(6):622–7.
5. Wakata S, Morioka S. Brain activity and the perception of self-agency while viewing a video of tool manipulation: a functional near-infrared spectroscopy study. Neuroreport. 2014;25(6):422–6.
6. Alphonso AL, et al. Use of a virtual integrated environment in prosthetic limb development and phantom limb pain. In Wiederhold BK, Riva G, editors. Annual review of cybertherapy and telemedicine 2012: Advanced technologies in the behavioral, social and neurosciences; 2012. p. 305–9.
7. Ramachandran VS, Rogersramachandran D, Cobb S. Touching the phantom limb. Nature. 1995;377(6549):489–90.
8. Fink GR, et al. The neural consequences of conflict between intention and the senses. Brain. 1999;122:497–512.
9. Bengson JJ, Kelley TA, Mangun GR. The neural correlates of volitional attention: a combined fMRI and ERP study. Hum Brain Mapp. 2015;36(7):2443–54.
10. Abboudi R, Glass A, Newby N, Flint J, Craelius W. A biomimetic controller for a multifinger prosthesis. IEEE Trans Rehabil Eng. 1999;7(2):121–30.
11. D'Alonzo M, Clemente F, Cipriani C. Vibrotactile stimulation promotes embodiment of an alien hand in amputees with phantom sensations. IEEE Trans Neural Syst Rehabil Eng. 2015;23(3):450–7.
12. Kuttiva M, Flint J, Burdea G, Craelius W. VIA: a virtual interface for the arm of upper-limb amputees. In: Second international workshop on virtual rehabilitation: IWVR; 2003, vol. 2. p. 119–26.
13. Antuvan CW, Ison M, Artemiadis P. Embedded human control of robots using myoelectric interfaces. IEEE Trans Neural Syst Rehabil Eng. 2014;22(4):820–7.
14. Kooijman CM, et al. Phantom pain and phantom sensations in upper limb amputees: an epidemiological study. Pain. 2000;87(1):33–41.
15. Willoch F, et al. Phantom limb pain in the human brain: unraveling neural circuitries of phantom limb sensations using positron emission tomography. Ann Neurol. 2000;48(6):842–9.
16. Powell MA, Kaliki RR, Thakor NV. User training for pattern recognition-based myoelectric prostheses: improving phantom limb movement consistency and distinguishability. IEEE Trans Neural Syst Rehabil Eng. 2014;22(3):522–32.
17. Koban L, Corradi-Dell'Acqua C, Vuilleumier P. Integration of error agency and representation of others' pain in the anterior insula. J Cogn Neurosci. 2013;25(2):258–72.
18. Ronchi R, et al. Right insular damage decreases heartbeat awareness and alters cardio-visual effects on bodily self-consciousness. Neuropsychologia. 2015;70:11–20.
19. Allen M, et al. Anterior insula coordinates hierarchical processing of tactile mismatch responses. Neuroimage. 2016;127:34–43.
20. Segerdahl AR, et al. The dorsal posterior insula subserves a fundamental role in human pain. Nature Neurosci. 2015;18(4):499.
21. Preissler S, Dietrich C, Blume KR, et al. Plasticity in the visual system is associated with prosthesis use in phantom limb pain. Front Hum Neurosci. 2013;7:311.
22. Casale R, Damiani C, Rosati V. Mirror therapy in the rehabilitation of lower-limb amputation are there any contraindications? Am J Phys Med Rehabil. 2009;88(10):837–42.
23. Ramachandran VS, Altschuler EL. The use of visual feedback, in particular mirror visual feedback, in restoring brain function. Brain. 2009;132:1693–710.
24. Diers M, Flor H. Phantom limb pain. Psychological treatment strategies. Schmerz. 2013;27(2):205–11.
25. Hellman RB, et al. A robot hand testbed designed for enhancing embodiment and functional neurorehabilitation of body schema in subjects with upper limb impairment or loss. Front Hum Neurosci. 2015;9:26.

26. Dietrich C, et al. Sensory feedback prosthesis reduces phantom limb pain: proof of a principle. Neurosci Lett. 2012;507(2):97–100.
27. Ortiz-Catalan M, et al. Treatment of phantom limb pain (PLP) based on augmented reality and gaming controlled by myoelectric pattern recognition: a case study of a chronic PLP patient. Front Neurosci. 2014;8:24.
28. Yanagisawa T, et al. Induced sensorimotor brain plasticity controls pain in phantom limb patients. Nat Commun. 2016;7:13209.
29. Yang DP, et al. Dexterous motion recognition for myoelectric control of multifunctional transradial prostheses. Adv Robot. 2014;28(22):1533–43.
30. Phillips SL, Craelius W. Residual kinetic imaging: a versatile interface for prosthetic control. Robotica. 2005;23:277–82.
31. Cipriani C, Controzzi M, Carrozza MC. The SmartHand transradial prosthesis. J Neuroeng Rehabil. 2011;8:29.
32. Castellini C, et al. Proceedings of the first workshop on Peripheral Machine Interfaces: going beyond traditional surface electromyography. Front Neurorobot. 2014;8:1–17.

Chapter 7
Prosthetic Control Systems

7.1 Introduction

Our ability to ambulate and manipulate within the environment is naturally provided by a neuromotor system (NMS) that has evolved over countless generations. The NMS can direct virtually infinite patterns of motions initiated by intentional commands from our brains. Most of these motions require little conscious attention, since the NMS has a built-in control system for directing the movement by means of closed-loop feedback, as described in this chapter. The main focus is on prosthetic control designs that can best replace the natural control mechanisms within the human NMS.

7.2 Artificial Control

A *control system is* a set of interconnected components whose task is maintaining a variable or set of variables within a specified range, even in the presence of external disturbances and noise. The classical example of an artificial control system is the temperature regulator, such as heating, ventilation, and air conditioner (HVAC). The three necessary components are (1) a thermostat for setting the desired temperature, (2) a source of energy, in terms of heat, ventilation, and cold, and (3) sensors to provide feedback on the actual temperature, as it is being modulated by the system. This control system regulates the variable, temperature.

There are two basic types of control systems: open loop and closed loop, as illustrated in Fig. 7.1.

The classical example of Open-loop control is a HVAC room heater operating in mode (a), which turns on a *fixed* power level (The Plant) when room temperature drops a certain amount below the thermostat setting, and turns off when the temperature reaches the set value. In closed-loop operation (b), the heater (Plant) emits

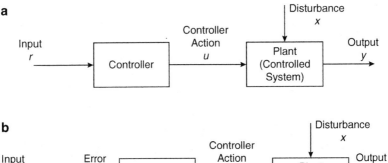

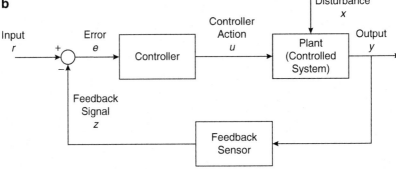

Fig. 7.1 Schematic of control systems. (**a**) Open loop, (**b**) closed loop

heat at a rate *proportional* to the difference between actual and set temperature. This quantity is the error between the set and actual value of the variable. Plant output diminishes as the error declines, shutting down when it becomes zero. The output is thus proportional to the error, unlike the open-loop system, which operates either on or off.

7.3 Natural Control of Limbs

While limb control is much more complex compared with an HVAC control system as depicted in Fig. 7.1, limbs are controlled by similar logic. Limb motions are characterized by three primary variables: joint torque, position, and acceleration. For the upper limb, these variables can represent the dynamics and kinematics of the hand reaching for a specific target, such as a plate, as shown in Fig. 7.2. Components of limb control are (1) a physical target, which is either a location at specific coordinates, or a force, (2) a source of energy, in the form of the neuromotor system, including muscles, and (3) sensors to monitor information about the actual position. The job of the controller is to minimize errors between the desired and the actual dynamics and kinematics, as the movement progresses. An example of a natural closed-loop control system is reaching toward a target, as depicted in Fig. 7.2. In this case, the positional difference between the hand and the desired object is the error signal which controls the movement. Human dexterity is based upon this control apparatus, which operates with minimal conscious supervision.

7.3 Natural Control of Limbs

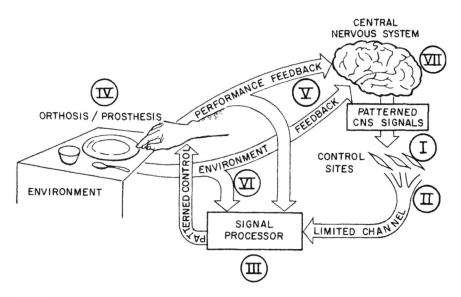

Fig. 7.2 Depiction of neuromotor control, as applied to a prosthesis or orthosis. https://encrypted-tbn0.gstatic.com/images?q=tbn%3AANd9GcSszFKa6rw4KNZbvu8SmuB7wmqSLfD1h64w0dg3igb4SFV-Z3GU&usqp=CAU

Limb motions are produced when the brain sends a "go" signal to the neuromotor system (NMS), which executes a pre-programmed sequence of muscle activations which are continually adjusting and correcting for trajectory errors. A network of sensors transmits the ongoing kinetic and dynamic parameters of the joint for sensory feedback and control. For reaching movements, there are several viable pathways and trajectories for the hand to reach a particular destination. While the NMS is typically pre-programmed, by learning, to generate the most desirable trajectory, in terms of efficiency and comfort, it is also endowed with a flexibility that can adjust parameters in real time, in the presence of obstacles or unexpected conditions. This optimal trajectory describes each point in space-time from starting point to the ending point at the target. Since no mechanism is perfect, however, there will be errors between the optimal and actual trajectory during the motion, for which the controller attempts to minimize. The NMS accomplishes this by monitoring the positional error in real time and attempting to correct the motion by minimizing the errors. The error is calculated simply as the Desired position—the Actual position. Motion is continually monitored with sensors, including proprioceptors, sensors of torques, and velocity, which send signals to the processor, in an attempt to correct the motion, in terms of acceleration and velocity. Prostheses use the closed-loop control system for best performance.

Closed-loop control can be incorporated into prostheses using sensors, processor, and actuators within the prosthesis, as shown in Fig. 7.1b. To illustrate the process, consider the example of reach for and grasping a plate. Here, the desired motion is positioning the limb correctly in space to land on the target. The variable is the position in space of the end effector, embodied either as the natural hand, or its replacement, the terminal device. Multiple muscles and joints must be coordinated.

An action sequence of prosthetic (or orthotic) arm manipulation is illustrated in Fig. 7.2: (1) a user's intention arises in the central nervous system (VII; brain and spinal cord), (2) signals (I) are sent through nerve pathways (II) to a signal processor (III), which produces appropriate signals to activate the muscles that operate the joints. These may be specific muscles or nerves within the spinal cord. In either case, the signals must be processed into a coherent message that tells the prosthesis to initiate reaching. Simultaneously with movement, sensors on the prosthesis send feedback signals about the environment to the signal processor (VI) and information about performance and the environment are sent (V) to the brain, in order to correct errors in trajectory, using negative feedback to reach a plate (IV), which is the desired target in the environment. The entire movement is sensed and monitored by the central nervous system, using feedback from both the environment (VI) and from the performance (V). These sensations about the ongoing motion help direct the arm accurately to the target. This monitoring and adjusting the motion represents closed-loop control. This controller follows Weiner's definition of feedback as "a method of controlling a system by reinserting into it the results of its past performance." All major components of a feedback control system are present in this system: (I) Signal Sources; (II) Transducers; (III) Signal Processors; (IV) Output Systems; (V) Feedback Receptors; and (VI) Local Feedback.

The biological components involved in limb movements are represented in Fig. 7.3. As seen, in response to an intent arising in the brain (1), a signal is sent to the spinal cord (2), which directs neuronal signals through peripheral motoneurons to specific muscles (3), which become activated via energetic mechanisms of sliding filaments (4).

> The motor cortex sends motor signals to the spinal cord neuronal network which sends its outputs to the muscles. The spinal cord combines motor signals with afferent feedback to generate the motoneuron outputs.

During arm movements, the natural control system sends sensory information from receptors within the joint to the spinal cord and higher centers within the CNS, as depicted in Fig. 7.4. Joint movement is detected by intramuscular sensors known as, "muscle spindles." These transmit neural pulses in proportion to their degree of stretch. Artificial sensors can be adapted for this purpose in prostheses, including strain gauges, piezo films, potentiometers, et cetera.

7.3 Natural Control of Limbs

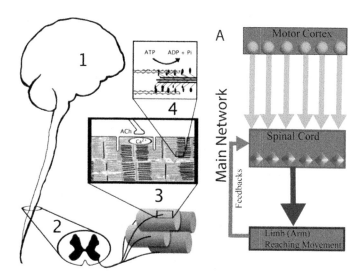

Fig. 7.3 (a) Control of arm reaching. https://doi.org/10.1371/journal.pone.0179288.g001. (b) Natural limb motion control. The four components shown constitute the natural limb motion control system. Here, muscle activation is initiated (1) by a signal from the motor cortex of the brain, (2) which travels via nerves to the spinal cord, (3) activating motoneurons. The muscle contracts via sarcolemmal depolarization, calcium release, and energy input (4), which finally produce muscle force. https://doi.org/10.1371/journal.pone.0056013.g001

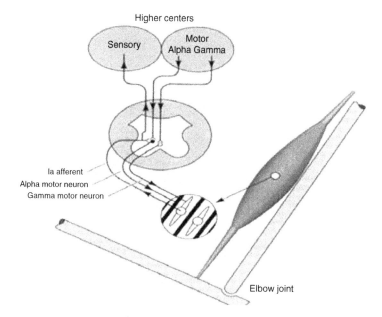

Fig. 7.4 Stretch reflex sensors at the elbow

7.4 Prosthetic Control

The prosthetic controller needs to direct the user's desired movements regardless of disturbances. Thus, the position, force, and accelerations of a prosthesis must be monitored and fed back to the processor, mimicking the natural system to the degree possible. Prostheses have incorporated components developed for robot arm control, such as servomechanisms for both actuation and trajectory tracking, accelerometers, and a variety of strain sensors. The schematics in Figs. 7.5 and 7.6 outline the basic components, inputs, and outputs of a prosthetic controller.

7.5 Adaptability of the Neuromotor Control System: Gravity Compensation

Proper functioning of both upper and lower limbs depends on their ability to move both with and against the ever-present force of gravity. Lifting and manipulation by the hands and swinging the leg for ambulation are fundamental requirements for activities of daily living. An important feature of the intact neuromotor system is its ability to adapt its output according to the vectorial relationship between gravity and limb motion. This phenomenon is seen by observing the dynamics and kinematics of arm motions during upward and downward swings, as depicted in Fig. 7.7a. This protocol measured arm trajectories of individuals while raising and lowering their arm 45° from neutral (straight out) at three different speeds. The data (Fig. 7.7b) show that the kinematics and dynamics of arm movements either lifting up or lifting

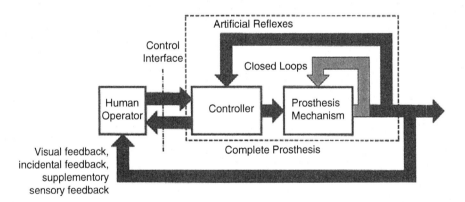

Fig. 7.5 A prosthetic controller. https://www.google.com/url?sa=i&source=images&cd=&cad=rja&uact=8&ved=2ahUKEwjfkMb2sLnmAhXOdd8KHSUmC68Qjhx6BAgBEAI&url=https%3A%2F%2Fwww.semanticscholar.org%2Fpaper%2FCHAPTER-32-DESIGN-OF-ARTIFICIAL-ARMS-AND-HANDS-FOR-Weir%2F1d5ef9e48b17f30cedd07cdd3f61f5474a80cb28%2Ffigure%2F28&psig=AOvVaw0-1nnGrKKegiJGvA9A5_kC&ust=1576558653754565. https://www.semanticscholar.org/ (Fig. 7.4)

7.5 Adaptability of the Neuromotor Control System: Gravity Compensation

down are quite similar, as shown in data plots of the arm motions in Fig. 7.7a. The motions were imaged and recorded at three different speeds, slow, "natural," and fast, and graphed to compare motion contours at different speeds, for both the upward and downward motions. The trajectories show that the position and velocity profiles with respect to time are nearly identical for both up and down motions, which is surprising, since motions are expected to be affected by gravity, and the gravity force vector is opposite for upward and downward motion. It is apparent that the motor control program compensates for gravity using continuous feedback of the kinematics and dynamics in a closed-loop configuration. The trajectories during raising the limb are essentially the same as those for lowering the limb. In other words, movement commands produce nearly the same motions regardless of

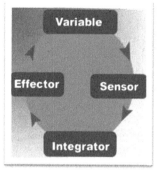

Fig. 7.6 Components of a controller

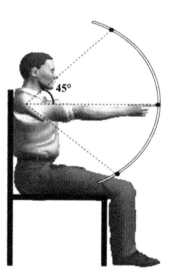

Fig. 7.7 (**a**) Arm dynamics and kinematics during both upward and downward motions at three different speeds. https://doi.org/10.1371/journal.pone.0022045.g002 Gaveau J, Papaxanthis C (2011) The Temporal Structure of Vertical Arm Movements. PLoS ONE 6(7): e22045. https://doi.org/10.1371/journal.pone.0022045. (**b**) Arm trajectories and dynamics during upward and downward movements. https://doi.org/10.1371/journal.pone.0099387.g001

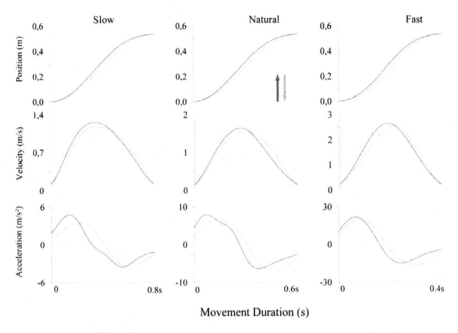

Fig. 7.7 (continued)

whether the arm is moving against or with the force of gravity. Thus since the kinetic and dynamic behaviors of the arms during upward and downward motions are virtually identical, it can be concluded that the intact neuromotor system compensates for gravity, using negative feedback of the motion, whether it is resisting or assisting the movement. Another observation from Fig. 7.7b is that the trajectory, velocity, and acceleration of the arm are each independent of both direction and speed and of motion, and show nearly identical curves. These features of the neuromotor control system endow humans with great dexterity, but have proven difficult to embody in artificial control systems.

7.6 Control Logic of Limb Function

The logic of closed-loop feedback of joint torque, which can apply to any limb movement, is illustrated in Fig. 7.8. Here, Box 1, "Excitation" represents the brain signaling a movement intent to the spinal cord. Box 2 represents the spinal cord, containing alpha motoneurons that are activated, and sending signals to activate the appropriate muscles (Box 3). Once the motoneurons signal the muscle, contraction begins, using stored and metabolic energy, as shown in Box 4a. Box 4 depicts the transduction of chemical energy into mechanical force, which produces a joint

7.6 Control Logic of Limb Function

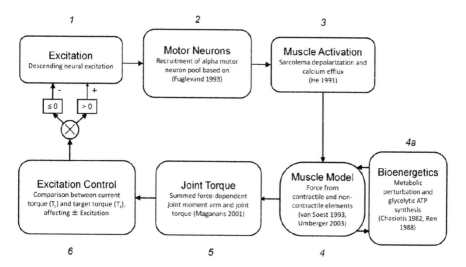

Fig. 7.8 Flow diagram of joint control. https://doi.org/10.1371/journal.pone.0059993.g001

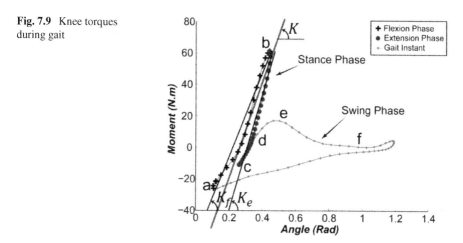

Fig. 7.9 Knee torques during gait

torque (Box 5). The magnitude of torque produced by muscular contractions is next compared with the target torque required. This torque error signal is used either to increase or decrease the neuromotor signal (Box 6). This system is designed to deliver the optimal torque to the correct joints in the proper sequence. Thus, natural coordination of leg joint torque control from the moment the heel strikes, to the moment the same leg leaves the ground, is illustrated, and readily reproduced as control logic. This same logic sequence applies to upper-limb motion as well.

Design of leg prostheses requires detailed knowledge of the motions and forces involved in normal gait. Typical torques experienced by the knee joint during two phases of gait are shown in Fig. 7.9. During stance phase, torque ranges from approximately −20 N-m at extension to about 60 N-m flexion just prior to toe-off.

The knee angle remains relatively straight in stance phase, as can be appreciated by noticing one's knee and the force while walking. During swing phase, however, the knee undergoes a large flexion. Considering that the lever arm of knee muscles is 10 cm or less, the flexion forces involved rise up to the 600 Newton range.

7.7 Gait: Controlled Falling

Most human movements on Earth are activities that work both with and against gravity. Any mechanical control system on Earth, whether natural or artificial, must operate in the presence of gravity. All human movements are affected by gravity, which plays both antagonistic and synergistic roles. The human motor control system exploits these features and can work both against and with gravity. The swing phase of gait, for example, involves a pendular motion, which works with gravity, as depicted in Figs. 7.10 and 7.11. This happens at the point at which the center of mass (CoM) is at its highest (Fig. 7.11) and represents partial recovery of the energy used to lifting the body during stance phase. This gait control logic is depicted in Fig. 7.5. Beginning the cycle at late stance phase for one leg, the phase moves to swing flexion when the foot load becomes less than a set threshold. Thus the neuromotor system is adapted to use muscles against gravity and recover a portion of the effort as inertia.

During walking the CoM rises and falls with each cycle, reaching the highest point at about 50–60% of the gait cycle, as shown in Fig. 7.11. At this point, the forward momentum of the body helps swing the free leg forward, and onto the ground, as if the body were momentarily in a free-fall state.

7.8 Walking Coordination

Walking is coordinated by several control systems, one of which is depicted in Fig. 7.12. Here we see that the muscle spindles within the *gastrocnemius* muscles of both legs sense their individual contractions and send this information via afferent

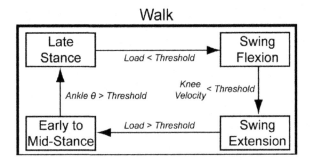

Fig. 7.10 Gait logic using a finite state machine. Basic components and logic of a feedback-controlled gait system are depicted

7.8 Walking Coordination

Fig. 7.11 CoM during gait. https://doi.org/10.1371/journal.pone.0117384.g001. John R. Rebula, Arthur D. Kuo

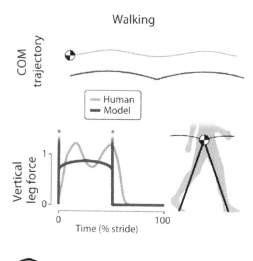

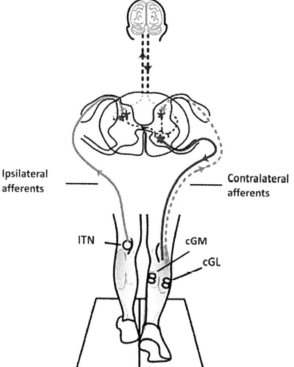

Fig. 7.12 Illustration of muscular coordination of the legs during gait. The *gastrocnemius medialis* (cGM) and *lateralis* (cGL) of the right leg are shown, sending afferent signals (dotted line) from muscle spindles. https://doi.org/10.1371/journal.pone.0168557.g001

nerves to the spinal cord. The cross-over arrangement of the sensory system as shown enables each leg to receive information about the opposite leg continuously during gait. This control system helps the two legs coordinate their motions during gait. This is important in the case when one leg, say the right, experiences an obstacle during gait that would require action by the left leg and foot, in order to maintain stability. Referring to Fig. 7.1, we see that the controlled system, or plant, in this case is the walker, whose stability is being controlled by the muscular feedback systems of both legs. The feedback sensors are the muscle spindles, which are known as "intrafusal" muscles, whose function is to sense the state of stretch of the "extrafusal" muscles which produce movement. The disturbance, x, is the amount of initial stretch produced by the tap to the knee. This produces a proportionate amount of stretch, y, which senses the amount of stretch and reports the magnitude back to the controller. The muscle spindles, like all neuromotor components, code the degree of stretch in a string of nerve impulses, whose frequency is proportional to stretch. Since neuronal output consists of an all-or-none action potential, the muscle spindle transduces the analog quantity of stretch into a series of pulses, which are analogous to a digitally coded signal, whose frequency is proportional to mechanical stretch. Thus, after muscle spindles encode the analog quantity of stretch into a digital format (z), they signal the reflex center in the spinal cord. The controller then activates the thigh muscle, contracting it to relax the stretch. This is a *negative feedback* system, since the initial disturbance (tap-induced stretch) leads to a controller action that *reduces* the effect of the disturbance.

Most joints of the human body employ controllers similar to the knee jerk reflex, for maintaining desired positions as well as to help direct the motion dynamics. For example, the reflex of the elbow shown in Fig. 7.3b is quite similar to those of other joints, such as the knee. Note that there are three types of neurons involved in this reflex, the 1a afferent, the alpha motoneuron, and the gamma motoneuron. The arrows denote the direction of nerve impulses from the neurons to the spinal cord, and into the brain.

For the unilateral amputee, leg coordination system described above is difficult to implement, due to the inadequacy of sensory feedback processing, and actuating the proper signals.

The standard muscle stretch reflex serves as a good example of negative feedback control. Consider the block diagram representation of this reflex, as shown in Fig. 7.13. Comparing this configuration with the general closed-loop control system of Fig. 7.1, one can see that the thigh muscle here corresponds to the plant or controlled system. The disturbance, x, is the amount of initial stretch produced by the tap to the knee. This produces a proportionate amount of stretch, y, *in* the muscle spindles, which act as the feedback sensor. The spindles translate this mechanical quantity into an increase in afferent neural traffic (z) sent back to the reflex center in the spinal cord, which corresponds to our controller. In turn, the controller action is an increase in efferent neural traffic («) directed back to the thigh muscle, which subsequently contracts in order to offset the initial stretch. Although this closed-

7.9 Tactile Feedback for Control

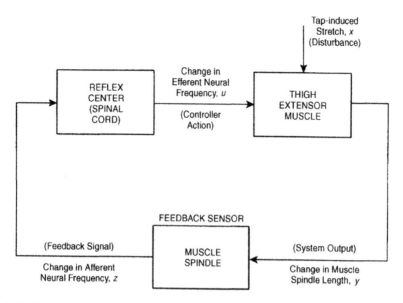

Fig. 7.13 Operation of knee jerk reflex

loop control system differs in some details from the canonical structure shown in Fig. 7.13, it is indeed a *negative feedback* system, since the initial disturbance (tap-induced stretch) leads to a controller action that is aimed at *reducing* the effect of the disturbance.

7.9 Tactile Feedback for Control

Loss of limb causes not only motor loss, but sensory loss as well, which is particularly acute for the upper limb. The lack of hand sensation is particularly devastating, because of its importance in feeling and manipulating things in the environment as well as in communicating physically with people. Many UL amputees abandon their prosthesis because it lacks sensation (Emily L. Graczyk, Anisha Gill, Dustin J. Tyler, Linda J. Resnik). Many stimulating devices have been developed in order to restore some sensory function to persons who have lost a hand or arm; however none have yet to restore a high degree of natural feeling. Direct re-connection of tactile information to the brain has thus far been unsuccessful; however, devices using sensory substitution are available. A sensory-enabled hand is illustrated in Fig. 7.14. This embodiment transduces tactile information with sensors on the fin-

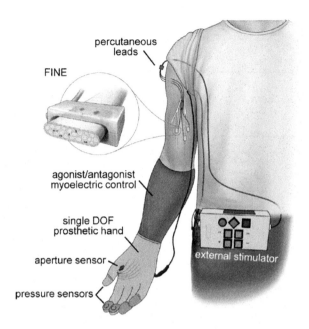

Fig. 7.14 Prosthetic upper limb with sensory feedback. https://doi.org/10.1371/journal.pone.0211469, Graczyk EL, Gill A, Tyler DJ, Resnik LJ (2019) The benefits of sensation on the experience of a hand: A qualitative case series. PLoS ONE 14

gers, into electrical pulses applied elsewhere on the anatomy, in this case, the upper arm. Sensations are represented in magnitude by the intensity of the electrical stimulation on the shoulder region, from percutaneous electrodes. The amputees who tested this device during daily activities for several days reported positive feelings about the sensory feedback, which gave them a better feeling of "ownership" and agency with the prosthesis.

7.10 Exercises

1. Make a first-order and a second-order control system. The quantity you should be controlling is viscosity of MRF (i.e., stiffness). Use reasonable values for stiffness, from published literature. Show how damping magnitude affects control and pick a reasonable value for it (not overdamped or underdamped).
2. Find and review two articles on the use of magneto rheological fluid (MRF, or other similar modality) in knee prostheses. State the mode of control and discuss the operation of the knee.

7.10 Exercises

3. Construct block diagrams to represent the major control mechanisms involved in controlling the knee. Clearly identify the physiological correlates of the controller, the plant, and the feedback element, as well as the controlling, controlled, and feedback variables. Describe how negative (or positive) feedback is achieved in each case. Show how the stance-control knee responds to gait in a patterned way, ensuring that the user does not fall.

Chapter 8
Limb-Prosthetic Interface

8.1 Introduction

Attaching hard prosthetic components to soft, and likely compromised, residual tissue is a major challenge to restoring limb function. The interface between user and prosthesis needs to be structurally strong, comfortable, and safe. The most common exoskeletal interface device is a plastic socket, into which the residuum fits, providing mechanical stability, a safe biochemical environment, postural correction, and functionality without compromising or damaging the soft and vulnerable tissue it surrounds. The structural connections between the residuum and prosthesis should be a natural transition between the prosthetic device and the body's native tissue, and become an extension of the limb. For optimal performance, the socket should have two-way communication with its user and adjust to ongoing conditions. Ideally, the user's operation of the prosthesis should be seamless and intuitive: it should replicate the natural limb function as closely as possible. The commonly available sockets are custom formed from plastic to enclose the residuum tightly but comfortably. Users often a wear a silicone sleeve on their residuum, to stabilize the connection, and reduce discomfort. In some cases, there is an endoskeletal option wherein the prosthesis is directly attached to the residual bone. This procedure is relatively rare, due to its surgical requirements, and is used only for special cases, as discussed later in this chapter.

8.2 The Socket Environment

Within the socket, residuum is subject to mechanical forces of pressure and shear, heat, chemical and physical irritation. This type of environment is generally not healthy for any part of the body, especially the residuum, which is vulnerable to tissue breakdown. The socket environment can be optimized by use of appropriate

© Springer Nature Switzerland AG 2022
W. Craelius, *Prosthetic Designs for Restoring Human Limb Function*,
https://doi.org/10.1007/978-3-030-31077-6_8

materials, design, and structural adaptations. The lower-limb socket applies large external loads encountered during ambulation to the residuum, which is vulnerable to damage. To minimize loads to sensitive areas, they must be directed to the most capable and healthy tissue, and especially to bony regions, such as the kneecap for BK amputees, and to the ischium for AK amputees, when possible. The upper-limb socket is simpler, since it does not typically experience body weight forces. The silicone sleeve helps to cushion and protect the limb tissue. A properly fitted socket should feel as an extension of the limb, akin to that experienced by tennis players wielding a racket.

Sockets are contoured to fit securely over the residuum with mechanical purchase primarily over the available bony surfaces, as detailed below. Socket materials need to be moldable or manufacturable into accurate shapes, compatible with soft tissue, and sufficiently strong to endure maximum expected stresses, while being light weight. Requirements include comfort, mechanical strength, flexibility, and safety.

8.3 LL Socket Fitting

Fitting the limb to the socket is crucial to successful restoration of mobility and function. An ill-fitting socket, no matter how advanced, will lead to discomfort, pain, and poor control of the mechanical components. Successful limb restoration requires close collaboration between the surgeon and prosthetist. Proper fitting of residuum within the socket is determined by custom shaping of both the socket and the residuum itself during surgery. Since each individual's anatomy, physiology, and lifestyle are unique, restoring the limb not only requires craftsmanship but also a thorough understanding of anatomy, physiology, and biomechanics. Another important factor affecting the socket interface is alignment of the prosthetic limb in relation to the residuum and the entire body, as described in Sect. 8.6.

8.4 Upper-Limb Sockets

The UL socket connects the terminal device (hand) to the residuum, where it needs to support all external forces without damaging the residuum. The socket is typically a cone-shaped plastic component whose shape and design depends on the level of loss, which ranges over the levels shown in Fig. 8.1. Socket length is determined by the length of the residuum, which ranges from loss of the entire limb below the elbow to a disarticulation of the hand, leaving the entire length of the forearm intact. The socket is form-fit to the user's residuum. The stability and strength of the socket depends on careful material selection. Common designs are plastic resins laminated with strong fibers, such as fiberglass or carbon. Strength of the sockets is calibrated by the type and density of fiber. To make a strong, yet comfortable attachment to the

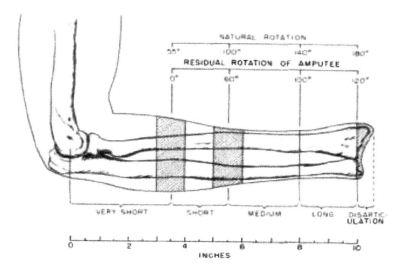

Fig. 8.1 Residual arm. From O&P Library

limb, sockets can be fabricated with a double wall, with the inner layer conforming to the stump and the outer wall providing structural integrity and cosmesis. UL sockets can be adorned with attractive patterns and texture. While many sockets are still made by hand, and with more primitive materials in less developed regions, newer materials and fabrication methods, including 3D printing, are revolutionizing socket manufacture.

The below-elbow (BE) socket should restore as completely as possible the function of the forearm, which includes opening of the hand by forearm flexion and hand rotation by means of pronation-supination. In the BE prosthesis, adequate forearm flexion is obtained rather easily, since the elbow and biceps are is intact. Wrist rotation, however, is limited due to the loss of forearm length, and subsequent limitation of its range of rotation. To achieve full wrist rotation, a rotatable prosthetic wrist is required, which requires control signals from the user. Control of the terminal device in space depends upon the socket to preserve residual elbow flexion and rotation of the BE residuum.

8.5 Fitting the BK Residuum

Most leg amputations are below knee (BK). A typical BK (right) residuum is depicted in Fig. 8.1. Note its irregular contours, scars, and apparent bony outlines. The anterior view of the leg shows the contours of the kneecap, which is the primary weight-bearing component.

Most BK amputees use the patellar-tendon-bearing (PTB) below-knee prosthesis, designed in 1958 by the Biomechanics Laboratory of the University of

California. The PTB prosthesis is designed to transmit most of the body weight to the patella, i.e., kneecap, by contouring a shelf at the top of the socket, where the knee can rest, as depicted in Figs. 8.1 and 8.2. The sides of the socket rise high above the knee to stabilize it against side loads. The PTB prosthesis uses the patellar tendon as a weight-bearing structure, since it can withstand high pressures due to its high stiffness. The socket material is plastic laminate that fits intimately over the entire residuum and is lined with a thin layer of soft material such as sponge rubber. Because the below-knee residuum is vulnerable to damage, socket shapes and features are designed to limit pressures on the residuum, and reduce friction, heating, and buildup of sweat, bacteria, and toxins. A conical shape of the socket helps distribute the body weight along the sides of the residuum, rather than its end. Often, the residual limb itself is surgically sculpted by the surgeon. This feature helps reduce edema and increase stability. Suspension of a PTB prosthesis is often supplemented by simple cuff, or strap, around the thigh just above the kneecap; sometimes a strap from the prosthesis to a belt around the waist is used. When suction alone is not strong enough to comfortably hold the limb in place, the harness, straps, belts, or sleeves become necessary (Fig. 8.3).

The prosthesis is donned by insertion of the residuum into the socket, which requires considerable force, since the air confined within the socket applies a resisting pressure. To allow the donning of the socket, an air valve is usually installed at the bottom of the socket, as shown in Fig. 8.4.

The residuum is usually covered with a tight silicone sleeve, as depicted in Fig. 8.5. A variety of sockets, fabricated with clear polypropylene, each having a pylon and foot installed are shown in the figure.

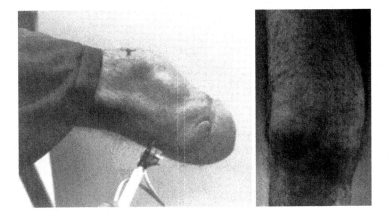

Fig. 8.2 Typical residuum of a below-knee amputation. Craelius lab

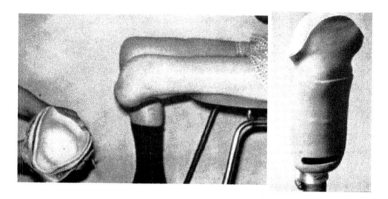

Fig. 8.3 Posterior view of a left BK being fit for a PTB prosthesis. The fit and alignment are adjusted for the individual. Profile of the PTB socket at right. Note the shelf for the patella tendon. *The Patellar-Tendon-Bearing Prosthesis for Below-Knee Criterial Amputees, a Review of Technique.* 1964. JAMES FOORT, M.A.Sc.2 O&P Library

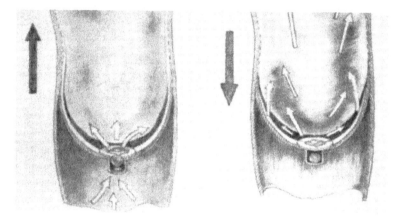

Fig. 8.4 Cycling of air pressure in socket. When the socket is in suspension, air is drawn into socket valve and stump sock. During weight bearing, air is expelled around proximal periphery of the socket. *Navy Prosthetics Research Laboratory*

8.6 Ground-Reaction Forces

The socket-limb interface must transmit all external forces, the largest of which are ground reaction forces (GRFs), to the skeleton without overloading the tissue or causing discomfort. GRFs are transmitted from the feet upward through the skeletal system. A person standing normally with both feet on the ground has a "base of support" (BoS) which transmits the GRF to the rest of the skeleton. The BoS normally includes the major plantar structures of the foot: *calcaneus* and *metatarsals, which are missing from an amputated leg*. For amputees, GRF transmits either through a prosthesis and one sound leg, or two prostheses, to the residuum. A lower-limb

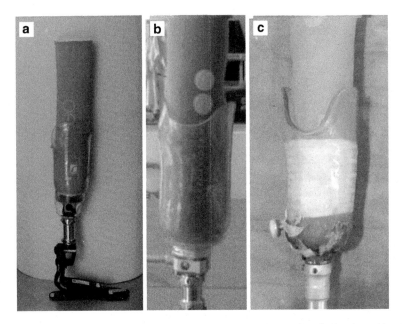

Fig. 8.5 Socket types. (**a**) Straight, (**b**) patella bearing, no valve, (**c**) patella bearing with valve. Valves are commonly used in sockets (Fig. 8.4), to allow air to be expelled from the socket during insertion of the residuum creating a vacuum. This feature helps in donning the prosthesis, and it also tightens the connection. doi:10.1371/journal.pone.0096988.g001 (Freely available)

prosthesis therefore, at minimum, must comfortably hold up the entire body weight. The force produced by body weight "pushes" down on the prosthesis and transmits through the prosthesis and onto the ground. The prosthetic socket must be strong enough to push up enough to hold this weight, and sockets must tolerate maximal expected loads, while remaining comfortable and safe.

Design criteria for limb sockets include not only the anatomy and physiology of the residuum but also the intact leg joints. Figure 8.6 shows a schematic of a BK leg in a socket and the arrangement of bones and tissues encased within a typical BK socket. Both the soft and hard tissues of the residuum are vulnerable to damage from the torques and moments caused during ambulation. The patella tendon, therefore, needs to carry the bulk of the GRF to minimize pressures on soft tissue. Also, socket fitting must be sufficiently tight to attach the limb with maximum security, but loose enough to limit tissue compression that could impair circulation and/or cause pistoning and skin friction (moving up and down), resulting in high air pressures within the socket (Fig. 8.7).

Use of the knee tendons for weight bearing is a highly successful adaptation. The patellar tendon is centrally located and branches out from the *quadriceps* tendon on each side. Like the patellar ligament, these tendons are also useful for weight bearing. To exploit these anatomical structures, an indentation in the socket wall between the lower edge of the patella and the tendinous insertion is made (See Figure 8.6). The tendon shelf is helps load bearing.

Fig. 8.6 Cutaway of a BK leg in a socket. The socket is secured with a waist belt, strap, and cast, and has a thick liner and shell (A)

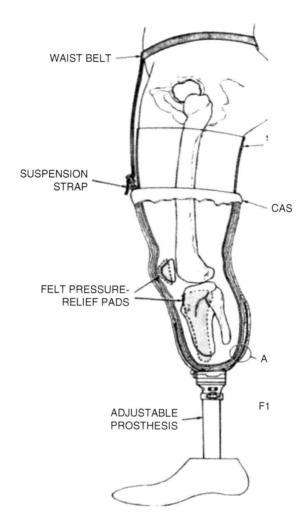

8.7 Installing the Leg Prosthesis

Restoring comfortable gait depends upon many factors. The prosthesis must be comfortable and be able to withstand the loads imposed upon it during all expected and possible extreme activities. The design process ideally begins prior to the surgery, with evaluation of the overall status of the patient, the limb metrics, and physical assessment of limbs. Following surgery, it is common practice to install an "immediate postoperative prosthesis," IPOP, on the residuum. This has many benefits, including reduction in swelling, quicker healing, and a psychological value in allowing the patient to see a complete limb, and begin to walk on it soon after surgery. In making the actual prosthesis, the prosthetist may choose among many different manufactured components, as well as design custom parts. As noted above,

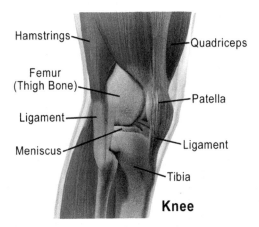

Fig. 8.7 Anterior view of right knee. At right, view of muscles and tendons of the knee. https://www.google.com/url?sa=i&source=images&cd=&cad=rja&uact=8&ved=2ahUKEwiNzbLb5rXmAhVDheAKHdmzB7cQjB16BAgBEAM&url=https%3A%2F%2Fen.wikipedia.org%2Fwiki%2FKnee&psig=AOvVaw0zJOkCkBV3zCAX6B9PsfCn&ust=1576435574347374. Creative Commons Attribution-Sharealike 3.0 Unported License. You are free to: Read and Print our articles and other media free of charge

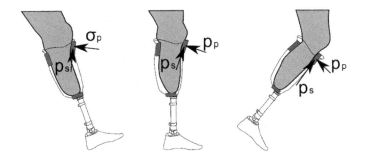

Fig. 8.9 Sensors, in red, are located on sensitive regions of the socket to register normal (p_p) and tangential (σs) stresses at the patellar tendon area and posterior of a trans-tibial stump during gait. Permission from Rights Link. *Medical Engineering & Physics*: Development and validation of a 3D-printed interfacial stress sensor for prosthetic applications P. Laszczaka, L. Jianga, D.L. Bader, D. Moser, S. Zahedi

the most customized component is the socket, which connects the user's residuum to the effector hardware. A soft liner, usually a silicone sleeve, is typically donned on the residuum, and the user might also wear a layer of one or more prosthetic socks to achieve a tighter fit.

The next step is attaching an adjustable leg (shank + foot) to the socket (Fig. 8.9), and allowing the user to stand, while the alignment is optimized with respect to ground. When both prosthetist and patient are satisfied with the alignment, a definitive pylon (shank) is attached. Pylons can be simple pipes or functional shock absorbing structures. The pylon is the major structural component and is usually a

8.7 Installing the Leg Prosthesis

metal or carbon-fiber composite tube that substitutes for leg bones. When these advanced materials are unavailable, other materials can be substituted, such as wood or plastic pipe. Pylons are sometimes enclosed by a cover, typically made from a foam-like material that can be shaped and colored to match the recipient's skin tone to give the prosthetic limb a more lifelike appearance. The foot prescribed is dependent on the client's functional level and needs. The most basic is the SACH (solid-ankle, cushion-heel) design but there are many advanced types, discussed in Chap. 4.

As seen in Fig. 8.8, an important determinant of prosthetic success is the alignment of the prosthesis to the residuum. During stance phase with the prosthetic leg alone, the support line must run medially to the support line, as shown at right. Misalignment is shown at left. Here the weight line of the body is not centered on along the load line, resulting in high pressure regions (arrows) and a gap in the socket.

Intra-socket forces during gait can be monitored with stress sensors as shown in Fig. 8.9. These sensors can be used for optimizing prosthetic shape, as well as for real-time adjustment of prosthetic hardware. Locations of the most pressure sensitive areas of the lower limb are depicted.

Particularly sensitive regions of the residuum are likely to suffer damage from friction and normal force. Tissue areas of pressure sensitivity vary along the leg, as shown in Fig. 8.11.

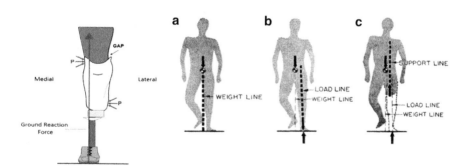

Fig. 8.8 Socket alignment. Force vectors supporting a leg. (**a**) The center of mass (COM) is the dot in the center of the body that represents where the body weight is accumulated. The weight line is the vertical line extending from the COM to the ground that the COM acts downward. (**b**) The load line is the line that represents the reaction force experienced between the foot and the ground. This reaction force includes both vertical and horizontal components dealing with loading and balance. (**c**) The support line is the line that the prosthetic is built to act upon supporting the body to maintain a balanced position; it should be coincident with the load line. Charles W. Radcliffe, O&P Library (no copyright). All O&P Library material is in the public domain. Contact Librarian: Charles King

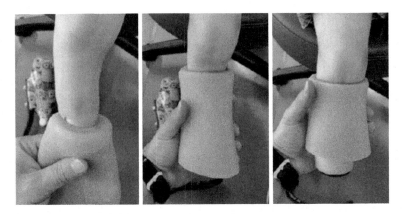

Fig. 8.11 Rolling a silicone sleeve onto residuum. Queensland Prosthetics, http://www.al1limb.com/

8.8 Socket-Skin Interface

Human integument (skin) is an amazing packaging material, yet to be reproduced artificially. Unlike inanimate wrapping material, skin regenerates itself, adapts to the local environment, and resists attack by all kinds of agents—chemical, physical, and biological. The features of living skin include: (1) respiration for exchange of oxygen and carbon dioxide; (2) regulation of body temperature by acting as a heat fin filled with sweat glands under control of the sympathetic nervous system; (3) a semi-permeable barrier for conservation of water and electrolytes; (4) sensors to report and record heat, cold, pain, and touch; (5) a conduit for nutrients, i.e., vitamins and hormones; (6) an active barrier against infection, and (7) regeneration capability. The skin renews itself by a continuous cycle of decay and growth. These functions operate well only if the skin receives proper care, exposure to air, freedom from constriction and irritation, prompt removal of waste products from its surface, and moderate temperature environments. The socket environment generally lacks optimal treatment of skin.

The socket environment is harsh: it is unventilated and tends to be hot, humid, and accumulates noxious chemicals and bacteria. The socket not only blocks the normal pathway for heat radiation, it promotes excess heat production due to extra energy expended for motions. Skin loses its ability to exchange heat, gases, and water (respiration and transpiration), but is continually exposed to the mechanical forces of pressure, shear, and friction. The risk for tissue breakdown is high, especially for those with compromised vascular circulation, secondary to diabetes, the most common source of leg amputation in the U.S. The soft tissue is tightly encased and compressed by the socket; these stresses are not desirable for tissue health, and need to be minimized by judicious shaping of the socket. Regardless of the socket design, the closed environment within, along with pressure and shear, is hazardous to the tissue. These hazards are present with all types of hard sockets, including the newer porous plastic laminates.

8.9 Forces on the Skin During Ambulation

Some friction against the skin is needed within the socket to support loads and prevent undesirable slippage; however its magnitude must be within tolerable limits within the socket. Too high friction causes high surface stresses and distortion during donning the prosthesis, as well as during ambulation. Too little friction, however, leads to slippage of residuum within the socket. A loose fitting socket results in slippage of the residuum, which may compromise stability; a very tight fit, in contrast, may offer a more stable connection, but increases the interface pressure leading to possible tissue breakdown. A medium level of friction thus must be low enough to minimize interfacial damage while meeting the requirements for secure attachment and control of the prosthesis. All materials can slide against skin, producing variable amounts of shear, depending on the friction coefficient. With very high-friction materials, slipping is minimal, but irritation and damage are still possible due to shear. Thus there are risks for both high and low friction materials, and selecting optimal features requires an individual approach for each user.

Shear is a force couple in which opposing tangential forces act on a body. In the socket, shear is caused by slippage between the subject's skin and the prosthetic socket, as well as by normal forces applied by the leg. This phenomenon can be seen as a tangential force, as shown in Fig. 8.10. The magnitude of shear during slippage depends on the amount of friction between the skin and the socket surface. Shear force increases with friction. Shear forces balance, since they always oppose each other. Evaluating the impact of shear on the limb involves knowing (1) the coefficient of friction of skin with various interface materials, (2) the magnitude of relative slip, and (3) the contribution of frictional shear to the load transfer. High friction between the skin and the socket tends to limit sliding of the socket against the skin, but the resulting shearing forces may damage the tissue. With too little friction, the socket will tend to slide up and down, which can also be harmful. There is thus a medium amount of friction for optimal performance.

As the residuum slides up and down in the socket during walking, the skin experiences various magnitudes of rubbing, wrinkling, and shear. Safe levels of these forces have been established using radiography and ultrasound on ambulating amputee subjects, of various weights. Using spiral X-ray computed tomography

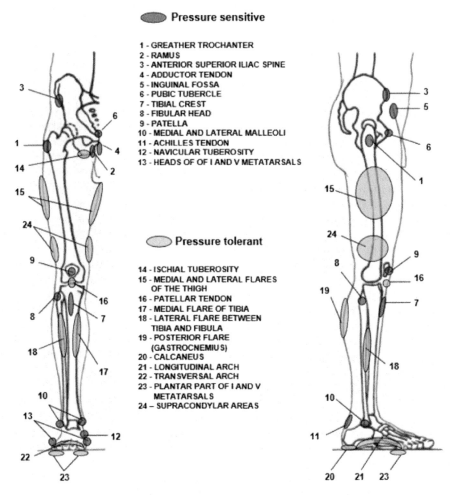

Fig. 8.10 Pressure sensitive areas on the leg. https://www.physio-pedia.com/images/8/81/Pressure_sensitive_and_tolerant_areas.png. The copyright holder has given special permission to use this work in Physiopedia. Biomechanics in prosthetic rehabilitation

imaging, estimates of limb slippage in the socket during axial static loading conditions increased from 2 to 6 mm when the applied load increased from 44.5 to 178 N. Slippage is as much as 22 mm for standard patella-tendon bearing (PTB) sockets and half that amount for PTB suction sockets. These magnitudes of slipping are not desirable and can cause irritation and ulceration.

8.10 Materials for Sockets

Skin shearing forces within the socket vary depending on the materials lining the socket. The common liner for sockets is silicone, a rubbery material, usually about 5 mm thick, that rolls onto the limb, much like a tight stocking. The friction coefficient, μ, between skin and silicone is very high, 0.9. The friction force, F, is calculated as $F = \mu N$. Since N, the normal force, against the skin is high, due to the tight silicone wrap, there is high friction, and correspondingly high shear, but little slippage between sleeve and skin. Friction can be reduced by reducing pressure in the socket by voiding air out the valve. There are risks to both high and low interfacial friction, and design considerations must be based on individual situations.

At the low end of the friction scale are materials such as nylon or Teflon, which have skin friction coefficients of 0.5 and 0.2, respectively. The coefficients of friction of human skin have been measured *in vivo* against five different prosthetic materials: aluminum, nylon, silicone, cotton sock, and Pelite. The measurements were made on six different anatomical sites of untreated skin using a *Skin Friction Meter* (M*easurement Technology, Inc.*). The average coefficient of friction was 0.46. The value was highest for silicone (0.61) and lowest for nylon (0.37).

For any prosthetic user, there may be an optimal balance between friction and sliding within the socket; however there are little data to guide socket design with regard to these forces. Finite element models of limbs under load have tested the effects of friction on load transfer from limb to socket and have shown an inverse relationship between friction and shear, but as friction declines, larger pressures are placed on the residuum to support the same load. The trade-off among the three factors of interfacial friction, tissue pressure, and socket stability thus necessitates an imperfect compromise for most users. The designer is faced with a decision between a tight socket fit with no slippage but high pressures and shear, and a loose fit with lots of slippage, either of which is potentially damaging to the user. The decision can be clarified based on the user's particular skin condition and overall preference. For some amputees, the safest solution would be the option of a direct endoskeletal attachment, discussed later in this chapter.

8.11 Protecting Tissue in the Socket

Trapping a residuum, which is often metabolically compromised, in an air-tight, hot, humid, environment, while subjecting the tissue to friction, shear, and pressure, is a recipe for tissue damage. Usually the primary tissue damage shows up as edema, caused by an imbalance in the interchange of materials between blood and body cells flowing through the capillary and lymphatic systems. Edema is commonly caused by uneven distribution of the socket compressive forces. When a relatively high *pressure* in one region squeezes fluid from venous and lymphatic vessels to a distal region of lower pressure, edema nevertheless can happen. Another pathway for edema is where muscles become paralyzed due to buildup of noxious chemicals, which compromises venous and/or lymphatic return.

The prime preventative remedy against edema is judiciously shaping the socket to distribute pressures and shear away from vulnerable regions of the residuum. This requires a thorough mapping of the residuum in terms of tissue integrity and shape. Socket walls can be sculpted to relieve pressures at bony prominences and tender areas, as shown in Fig. 8.11. An air valve at the bottom of socket also relieves pressure. With a vacuum inside the socket, the residuum is held tightly, with a level of compression that oscillates at each step, ideally delivering gentle external pressure pulses to aid the return of blood and fluids to the venous system. Such gentle pulsations (<40 mmHg) can benefit circulation.

8.12 Material Selection for Structures

Early versions of lower-limb prostheses were crafted from willow wood, a sturdy and easily machined wood. Today's materials include aluminum, steel, titanium, fibers of carbon and glass, and various plastic resins. In third world countries artificial legs are commonly made with locally available materials, such as wood, PVC pipe, and leather. Many new materials are being deployed in prosthetic interfaces. For hip and knee implants, a polyethylene formulation *Durasul* has shown no measurable wear after 20 million cycles in a hip simulator. This material may increase the lifetime of hip and other implants. A resource to help engineers select materials and methods for preventing tissue damage is "Wounds International Enterprise House" www.woundsinternational.com.

> Pressure, shear, friction and microclimate in context. PUBLISHED BY: Wounds International Enterprise House 1–2 Hatfields London SE1 9PG, UK. Tel: + 44 (0)20 7627 1510 Fax: +44 (0)20 7627 1570 info@woundsinternational.com, www.woundsinternational.com

8.13 Laminates

Modern sockets are made of plastic laminates, which are lightweight, easy to clean, and compatible with human tissue. They are easily molded over plaster molds of the residuum, making an accurate negative image of the residuum. Laminates can be made with ranges of stiffness: extremely rigid or for maximum strength, or with degrees of flexibility for comfort. Plastic laminates are easy to mold for close fitting to the limb and provide strong foundation for the attached hardware. Drawbacks are the plastic used is not porous to water or vapor, so it traps heat and moisture.

Laminates, however, can be made porous during curing, at the expense of some strength. Carbon fiber can be integrated into the laminate to boost strength, with minimal weight. The other drawback of porosity is the tendency for biological and other materials to accumulate in the pores, which require cleaning. Most of the common mechanical parts, made of steel or aluminum, attach well to the socket.

Plastic laminates are composites fabricated by applying a resin (matrix) to lay-ups of one or more layers of materials (fibers) on a plaster mold of the residuum. By adding a small amount of strong fiber to a weak resin, laminates form a strong but inexpensive and light weight structure. Strength of a laminate material is determined primarily by the material properties of the fiber and the degree of intermingling or penetration with resin that is achieved; strength (both tensile and bending modes) is always greatest along the axis of the fibers and weak in the orthogonal direction. For this reason, laminates are usually made using a fiber weave, along 2 or more axes. In the absence of weaved fiber, which is expensive, the fiber axes can be changed from layer to successive layer to endow the material with stiffness in all directions. If a lay-up of resin (composite laminate) were to contain only a single fiber axis or weaves which had vertical orientation, the vertical axis would be strong, but the orthogonal axis would be very weak.

8.14 Material Properties of Selected Prosthetic Laminates

The general rule for laminates is that the modulus increases as the fraction of fiber, as:

$$E_c = E_M(1-V_f) + E_f V_f$$

where E_c is the composite modulus along the axis of fibers, E_m and E_f are the matrix and fiber moduli respectively, and V_f is the volume fraction of fiber.

Physical properties of individual layers in the lamination lay-up are important because they help determine the physical properties of the entire lamination. The amount and type of fiber used in the laminate determine to a large degree its strength and stiffness, as shown in Fig. 8.12.

Strengths of selected materials tested are presented in Fig. 8.10. Each of the eight different fibers was mixed with three different resins, as shown. The resins used, Epoxyacrylate, OttoBock, and IPOS, are all commercial products. The relative strengths of carbon, fiberglass, and Nyglass laminates are approximately 100:50:1, respectively. Both carbon and fiberglass are very strong, but they have drawbacks as prosthetic socket materials; due to their brittleness, they break rather than stretch during post-fabrication modifications, which reduces strength significantly. Moreover, these fibers break easily when holes or cutouts are made in the socket, causing edge stress that weakens the structure. Combining different fibers also has drawbacks, since such hybrids can be weaker compared with a single fiber laminate.

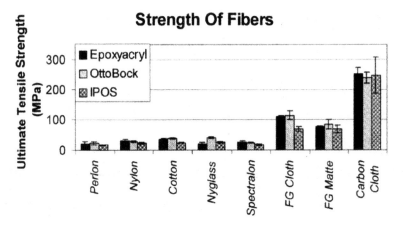

Fig. 8.12 Strengths of selected laminates are compared. Each fiber is embedded in one of three resins, as shown (Phillips & Craelius)

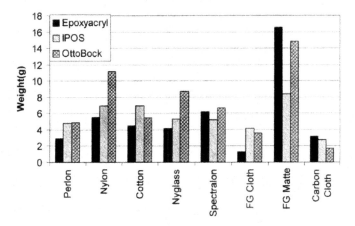

Fig. 8.13 Weights of different laminates. P Phillips, Craelius; JPO Journal of Prosthetics and Orthotics Volume 17 • Number 1 • 2005

Note that the modulus of the SPT resin (matrix) is roughly equal to that of nylon. In general, the material properties of prosthetic laminations are not well documented, which is a prerequisite for fabricating lighter and more cost-efficient prosthetic sockets, without sacrificing safety.

Prosthetic laminates consist of a fibrous material embedded into a resin. Strengths and weights of various combinations of resins and fibers are shown in Figs. 8.10 and 8.11.

Comparing strength-to weigh ratios (Figs. 8.12 and 8.13) for the various compositions, it is seen laminates with fiberglass and carbon cloth are considerably stronger than all the others. Also, the strength-to weight ratio for carbon laminates is much greater than all the other laminates (Fig. 8.11).

8.15 Dynamic Strengths of Laminates

Strength comparisons of two common laminates, fiberglass, and SPectralon (OttoBock) are shown in Fig. 8.11. Note the high relative stiffness of fiberglass cloth and its abrupt fracture at a low strain. SPectralon, in contrast, exhibited low stiffness and a relatively large strain. These results may indicate that Spectralon is a safer material for sockets; since it is not stiff, it is highly ductile, in comparison with fiberglass, which fails abruptly, a very unfavorable event, should it happen in action (Fig. 8.14).

8.16 Preserving the Residuum

From a user's point of view, the foremost requirements of a limb prosthesis are likely to be comfort, function, and appearance. While each user has different priorities, fitting the socket securely and comfortably is the most important goal of the prosthetist, because discomfort implies defects that may lead to skin lesions and tissue breakdown, and likely discarding of the prosthesis. Establishing a healthy interface between the residuum and the socket is therefore critical to prosthetic design. At a minimum, the interface, particularly for the lower limb, must be tight enough to maintain contact and bear weight, while not causing constriction and tissue damage. Ideally, the socket interface should promote tissue health by permitting tissue respiration and stimulating blood flow.

As noted earlier, the tissues of the residual limb are compromised in many ways and not well adapted to load bearing, friction, and shear. The large forces associated with the impact at heel strike, in addition to other phases, must be transmitted through the soft tissues and ultimately through the skeletal and musculotendinous structures. The user's perception of discomfort with the socket likely relates to excessive loading of the residual limb soft tissues or other parts of the residuum.

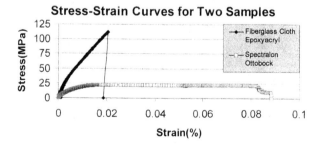

Fig. 8.14 Stiffness and strength comparison of two socket materials. Materials consisted of a plastic resin impregnated with either fiberglass cloth and Spectralon, a material made by OtoBock. (From Craelius Lab) (Phillips& Craelius), JPO Journal of Prosthetics and Orthotics Volume 17 • Number 1 • 2005

Socket materials are designed for mechanical flexibility in bending, expanding, and contracting along with massaging rather than irritates the residual limb, while maintaining total contact. The socket should promote venous return and tissue circulation during ambulation. This feature is especially important for the dysvascular and/or older amputee, who is otherwise subject to swelling and pain. During each gait cycle, the socket should expand and contract as the residual muscles expand and contract at foot contact. During swing phase, the socket contracts as the muscles relax and clamps back onto the residual limb.

8.17 Custom Designing and Fitting Limb Sockets

Limb sockets, like all prosthetic components, have evolved from a craft by amateurs into a science practiced by highly trained professionals, using high technology. Socket design is essentially a tight fitting hard sock for the residual limb. Manufacturing the socket can be done with either analog or digital methods. The analog method involves making measurements of the residuum with tape, making plaster casts of the patient's limb and measuring and annotating them with notes of critical landmarks. The plastic socket is then molded plaster cast. In some cases, the socket is molded directly onto the limb, using special resin that does not overheat.

Today's imaging technology has major impact on socket design and manufacturing. Digital scanners, as shown in Fig. 8.15, can precisely measure the limb dimensions, whether wearing a sleeve or not. These data can be used to print out a custom socket. Digital methods begin with scanning and digitizing the limb's anatomy and texture data which are input to one of the many CAD and photogrammetry apps available for creating 3D digital models. Tools such as cell phone cameras and scanners, as shown in Fig. 8.16, are employed. With the digital model, a

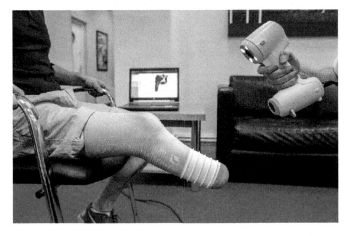

Fig. 8.15 Digital Scanning of prosthetic sock. Courtesy of Ability Matters, Abingdon, Oxfordshire, UK

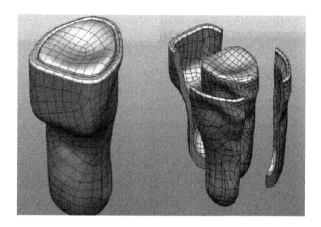

Fig. 8.16 Limb scanning for Prosthetic Fitting. CAD model of BK Limb and Socket. Procedia CIRP 60 (2017) 476–481. This is an open-access article under the CC BY-NC-ND license (http://creativecommons.org/licenses/by-nc-nd/4.0/). Vimal Dhokiaa, James Bilzon\ Elena Seminati\ David Canepa Talamas\Matthew Young\William Mitchel. doi: I 0.1016/j.procir.2017.02.049

personalized liner and socket can be fabricated and manufactured. These technologies are rapidly advancing and are producing personalized interfaces with custom components. Several companies offer socket manufacturing, including *High-Fidelity Interface System (HiFi; BioDesigns, Inc, Westlake Village, CA), Marlo Anatomic Socket*, and *Ability Matters, as shown in* Fig. 8.13. The Northwestern University *Flexible Sub-ischial Vacuum (NU-FlexSIV)* socket uses active vacuum suspension in an attempt to enhance reliability of suspension while maintaining the benefits of ROM and comfort.

These features help to stabilize the prosthesis, absorb shocks, and promote circulation in the residuum.

Skin breakdown and limb pain often experienced by those using lower-limb prostheses with sockets are detrimental to mobility. Those who experience severe socket-related problems may have the option of being fitted with an endoskeletal limb attachment, which fuses the pylon directly to the bone. This procedure has been successful, but it requires a long period for bone growth to solidify around the prosthesis. The attachment to bone requires many months, after which the user may be able to ambulate (Hagberg and Branemark 2001).

Attaching the prosthesis directly to the residual bone bypasses the need for a socket and ideally should provide more natural mobility. The direct mechanical connection between the limb and prosthesis can provide better control, with more natural sensations delivered directly to the limb. Surgery for endoskeletal restoration is complicated, requiring a lengthy recovery period to allow solid integration between the bone and the pylon. Examples of endoskeletal restorations are shown in Fig. 8.17.

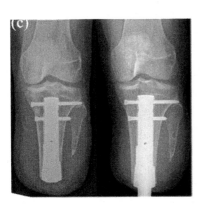  

Fig. 8.17 Examples of patient-specific implants. (Left) Postoperative (Right) integration into the bone demonstrated after 1 year. Clinical Rehabilitation 2019, Vol. 33(3) 450–464 © The Author(s) 2018 Ruud A Leijendekkers, Gerben van Hinte, Jan Paul Frölke, Hendrik van de Meent, Femke Atsma, Maria WG Nijhuis-van der Sanden and Thomas J Hoogeboom

8.18 Summary

While socket designs, fabrication, and monitoring have undergone major improvements by exploiting technology, there is still much room for improvement. Many LL amputees are expecting to engage in more normal and increasingly vigorous activities such as running, climbing, dancing, etc. This is because successful new designs in prosthetic hardware such as knees and ankles have opened more functionality possibilities for amputees, even exceeding those of natural joints. The socket requirements therefore continue to increase. Extreme activities, beyond just walking, cause transmission of large forces to the limb, including shear, compression, and torsion, which are risking damage to the residuum. Newer sockets provide total-surface contact for more even distribution of pressures. Air cushioning is incorporated to lessen the impacts. More intelligent socket designs are a major endeavor for the prosthetic industry.

8.19 Endoskeletal

Terminology

Stress	The amount of force applied over a given area
Strain	The amount of deformation for a given length
Young's modulus	Stress/strain: the amount of force to obtain a certain amount of deformation Synonymous with stiffness

Ultimate tensile strength (UTS)	Maximum force applied before a fiber breaks
Yield strength	Maximum force applied before permanent deformation occurs
Bending stiffness	Modulus during bending
Brittleness/ductility	Relative terms describing how much plastic deformation occurs before fracture; brittle = small deformation, ductile = high deformation

8.20 Exercises

1. Look up the "specs" of one of the devices for measuring intra-socket pressure. What range and resolution (both magnitude and frequency or response time) does this measurement have?
2. If SPT resin were combined with carbon fibers at 5% volume, what would the change in modulus be? Compute the same for fiberglass and compare (qualitatively) the brittleness of the two composites.
3. Define the difference between stress relaxation and creep and show formulae for both.
4. Calculate the critical bending moment for shear failure of a typical aluminum pylon and compare it with titanium.
5. Explain the difference between hysteresis and viscoelastic behavior. Give an example of each, with specific materials.
6. Complete the model from the workshop.

Bibliography

1. Mitchell CL. O&P Library. Artif Limbs. 1963;7(2):1–42. (Adapted from Mak et al. and a variety of sources).
2. Ohio. Ohio Willow Wood Company. 2017 [cited 2017]. Available from https://www.willowwoodco.com.
3. National Academies. NAKFI: smart prosthetics: exploring assistive devices for the body and mind: Task Group C, November 9-11, 2006, The National Academies in Conference, Arnold and Mabel Beckman Center of the National Academies. Irvine: The National Academies; 2006.
4. Mak AFT, Zhang M, Boone DA. State-of-the-art research in lower-limb prosthetic biomechanics-socket interface: A review. J Rehabil Res Dev. 2001;38(2):161–73.

Chapter 9
Energetics of Ambulation

9.1 Introduction

Lower-limb prostheses must be custom designed for the individual, whose gait pattern is unique. Gait restoration via prosthesis should approximate as closely as possible the natural gait with minimal training and effort from the user. A primary factor for success is based on the energy required for ambulation, which is complex and not fully understood. Differences among potential users are not only body morphometry and age, but their occupation, location, activity level, style of movements, and favorite sports or recreation. The designer must consider these factors to optimally restore ambulation. While technology has vastly improved our understanding of gait and how to restore it, restoring natural human gait is not yet exact. One of the reasons for this is because most gait data we currently have were obtained from users ambulating in steady-state conditions in a laboratory, which are not representative of the real world. This chapter outlines methods for maximizing users' gait, by optimizing energy usage, and exploiting body movements for powering electronic components.

9.2 Natural Energy Sources for Ambulation

Normal ambulation is powered by two sources: skeletal muscles and gravity. The muscles are fueled by (1) internally stored energy and (2) renewable metabolic energy. Gravity contributes both positively and negatively to energy, at different phases of gait. Efficient movement of limbs involves legs swinging as pendula, using gravitational energy to power the gait. Efficient gait uses gravity judiciously to minimize energy expenditure, as discussed in this chapter.

9.3 Cost of Transport (COT)

The energy used in ambulation can be accurately measured with biomechanical tools. The fundamental variable relating energy usage is "Cost of Transport" (COT), which represents just the extra energy used for ambulation, over and above that used for the basal (at rest) energy of operating the body's internal machinery. During ambulation, the head, arms, and torso, the "HAT," are propelled forward progressively and vertically in a rhythmic manner for each gait cycle. The alternating pendular action of the legs and arms in normal gait conserves energy by storing it as the body rises during stance phase, and returning a portion of it during the swing phase, as the body "falls" down to the next step. In other words, work is done in the rising phase, and part of that work is returned as energy in the falling phase. Normal ambulation thus involves the rhythmic rise and fall of the center of mass (COM), as depicted in Fig. 9.1a.

9.4 Artificial Energy Sources for Ambulation

Advanced prostheses employ actively powered joints, including hip, knee, and ankle for the lower limb and elbow, wrist, and hand for upper limb. These devices are powered artificially, usually by battery. These advanced prostheses can restore near-normal ambulation and better functionality to the upper limb; however, the batteries add weight and are subject to failure during operation. For this reason, energy harvesting is a solution based on exploiting body motions for re-charging the power source, discussed in Sect. 9.1.

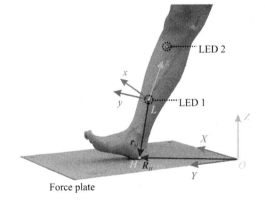

Fig. 9.1 (**a**) Measuring foot strike and motion of leg. https://doi.org/10.1371/journal.pone.0197428.g001

Force plate

9.5 Measuring Energy of Gait

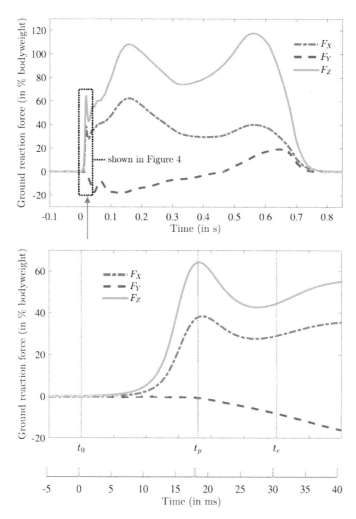

Fig. 9.2 (**b**) Ground reaction forces in the X, Y, and Z (mediolateral, anterior-posterior and to vertical) axes, as shown in Fig. 9.1. (**c**) Expanded record (40 ms) of the GRFs shown in Fig. 9.2a

9.5 Measuring Energy of Gait

A primary tool for analyzing gait efficiency is the force plate, shown in Fig. 10.1. This records the force in three axes as the foot strikes the ground and can also track leg motion using LEDs and a camera. The most important force for energy considerations is the vertical ground reaction force (GRF), measured here along the "Z" axis.

Forces during stance phase of one leg are quantified as percentage of subject's body weight (in this case, 727 N) and plotted during heel strike-to-stance phase. The impulse force at heel strike is outlined by the dotted rectangle and shown in expanded view in Fig. 9.2b.

In Fig. 9.2b we see the vertical GRF shortly after heel strike exceeds 100% of body weight, and lightens during weight acceptance, and again exceeds body weight at the toe-off. Figure 9.2c shows an expanded view (first 40 ms) of Fig. 9.2a. The peaks occurring between 15 and 20 ms are shown within the box (dotted lines) of Fig. 9.2b.

9.6 Energetic Contributors to COT

During normal gait, most of the COT is expended in lifting the CoM, as shown in Fig. 9.3. Here, the COM trajectory traces show the human data on top, and a "stick" model below, peaks during single stance, and falls to a minimum at double stance. The ground reaction vertical force shows a peak at the initial contact and a second peak at toe-off and a track of the COM (center of mass) during gait.

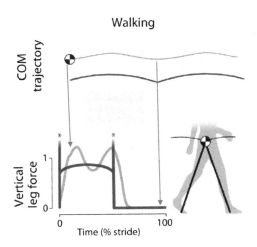

Fig. 9.3 Human COM trajectory & stick model COM trajectory & vertical Force. https://doi.org/10.1371/journal.pone.0117384.g001. Vertical ground reaction force and COM during a single stride. Gray traces of vertical leg force show the GRF of a human subject during one stride (i.e., heel strike-to-heel strike). The two peaks represent heel strike at left and toe-off at right. Left asterisk shows heel strike, and right asterisk shows toe-off. Vertical force is an impulse (vertical lines) at the heel strike and toe-off. Note the CoM trajectory is lowest at heel strike and toe-off and maximum in mid-swing. This reflects the fact that the body lifts up against gravity during single stance phase

9.6 Energetic Contributors to COT

Figure 9.3 contains a wealth of information about gait. First, at top are shown two traces of the COM during one stride. Top trace is human data and bottom trace shows the COM of a "stick leg" model, which has high stiffness. For this reason, the GRF impulse is essentially infinite. The COM trajectory of the stick leg also shows a sharp discontinuity at midpoint, which is wasteful of energy, and undesirable. This illustrates the need for a pliable limb.

The measure of energy conservation is the proportion of potential energy that is used in raising the COM that is converted into kinetic energy. Complete exchange between potential and kinetic energy occurs when the two energies are always moving in antiphase. The most direct comparison to walking is the motion of the legs that move like pendula.

Figure 9.4 shows the point of contact of the heel with the ground as a function of time, $H(t)$, and a measure of the foot-ankle deformation over time, $S(t)$. GRF was measured using a piezoelectric Kistler force plate, with a sampling frequency of 3000 Hz. Deformation due to heel strike is primarily attributed to the soft tissue in the heel pad.

A photograph (Fig. 9.4) of a human heel pad contacting the ground shows its deformation and that of the ankle. The global instantaneous position of the heel point $H(t_o)$, defined in the global coordinate system XYZ as shown in Fig. 9.1, is at the instant of initial heel contact at time t_o during heel strike. Due to deformation of the heel pad, foot, and ankle, the heel point $H(t)$ moves over the ground during heel strike over the ground surface, S. The movement of the heel point during heel strike is a measure for the foot-ankle deformation $S(t) = RH(t) - RH(t_0)$. The work dissipated by the foot-ankle is the surface integral of the force, as in Eq. (9.1).

$$W = \int_{S(t_0)}^{S(t_e)} F \cdot dS$$
$$= W_X + W_Y + W_Z$$

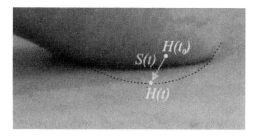

Fig. 9.4 Photograph of human heel making ground contact, with vectors H and S representing points of heel contact and foot deformations both as functions of time as the heel progresses on the ground. John R. Rebula, Arthur D. Kuo PLOS ONE I https://doi.org/10.1371/journal.pone.0117384

9.7 Ambulatory Efficiency

The key to energy-efficient gait is maintaining inertia: the sum of inertias of the HAT and the locomotory apparatus; these should progress smoothly, without accelerations and decelerations. This can be achieved by properly coordinated muscular activity. Energy efficiency requires conservation of inertial energy within each step. For example, the COM must keep moving forward smoothly to preserve inertia; a pause at mid-stance will cancel the inertia. Note from Fig. 9.3 that as energy is expended during the rising of stance phase, a portion of that energy is returned by gravity in the falling phase. Ideal gait can thus be characterized as controlled falling.

While COT is heavily influenced by the vertical movement of the COM, other factors affect it, including, walking speed, the ratio between walking speed and frequency, and step width. For example, fast walking with short rapid steps is not efficient, and step width that is either too wide or too narrow will cause excess muscular activity.

COT can be quantified by instrumented gait analyses that measure joint moments and power and muscle forces.

Energy considerations are essential to designing advanced LL prostheses, because gait can consume a large amount of both metabolic energy and battery energy, since they run on both body power and external power sources. Loss of power during ambulation is dangerous, since it can mean locked knee or ankle, leading to falling. Moreover, energy is needed to tune the prosthetic motor control according to the phases of gait, similarly to that of natural muscles that are tuned to minimize COT.

Gait speed is the product of stride frequency by stride length and is strongly affected by gait speed in humans: it is minimized at intermediate walking speeds of 4.5–5.4 km/h (1.25–1.5 m/s) and rises rapidly as speed increases above or decreases below this optimum range. Many walkers naturally develop an optimal speed for efficient energy usage, as shown in Fig. 9.6. With regard to running, the work done on the center of mass during a unit distance decreases with speed; however this gain is offset by the work done to oscillate the segments of the body relative to each other, i.e., internal work, which increases with speed. During walking, the pendular transfer of kinetic and potential energy of the legs is greatest at intermediate speeds, reducing the total mechanical work that must be performed by muscles at these speeds. The high external work at low running speeds and the high internal work at high running speeds might produce a U-shaped COT relationship. Thus, the observed energetically optimal walking and running speeds are consistent with the contractile physiology of skeletal muscle and biomechanics of terrestrial locomotion.

The key muscles involved in leg swing during ambulation are shown in Fig. 9.5. Energy efficiency is dependent upon the ability of these muscles to swing the leg like a pendulum. In the case of a prosthetic knee, this must be programmed to flex at the proper phase in order to maximize energy gain.

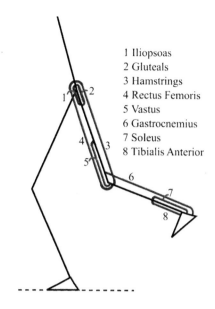

Fig. 9.5 Muscles involved in leg swing during gait. https://doi.org/10.1371/journal.pone.0222037.g001. Anne D. Koelewijn, Dieter Heinrich, Antonie J. van den Bogert

1 Iliopsoas
2 Gluteals
3 Hamstrings
4 Rectus Femoris
5 Vastus
6 Gastrocnemius
7 Soleus
8 Tibialis Anterior

9.8 Energy-Efficient Walking Speeds

9.8.1 Walking, Power, and Work Units

The amount of energy required for walking can be measured by the amount of oxygen consumed per kilogram body weight per unit distance traveled (mL/kg-m). The O_2 cost is determined by dividing the power requirement (rate of energy expenditure) by the speed of walking. Gait deficiency is assessed. Gait efficiency is measured by comparing the energy cost of a particular gait to the norm value for normal walking (Fig. 9.6).

The data taken from a subject walking show the effect of walking speeds on the ICOT, based on a walking model. It shows that the preferred walking speed is slightly more energy efficient than a speed of 130% of the preferred walking speed. Walking at slower speeds (70% of preferred speed), however, is more costly.

Optimum speeds during running vary widely among runners, and no general trends are apparent. This is mostly due to the fact that an energetically optimum speed is mostly an individual trait, and the group is quite heterogeneous. Estimating the energetically optimal speed for a specific runner is best done by measuring O_2 consumption at a range of speeds.

Compared with most species of mammals, humans are near the most economical bipedal walkers and possibly long-distance runners. In comparison with great apes, humans are economical walkers, probably because of our body size, long legs, and structural capacity to store and recover elastic strain energy.

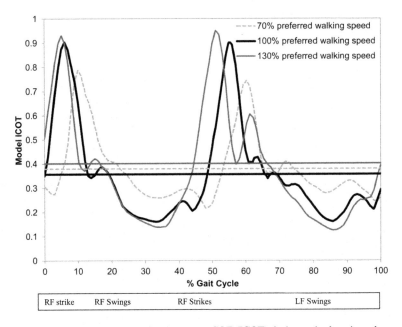

Fig. 9.6 Role of walking speeds on instantaneous COT (ICOT) during a single gait cycle

9.9 Walking Energy Harvesting

While gait requires energy input, it can also produce energy for operating electrical mechanisms and processors within the prosthesis. Advanced prostheses use sensors, actuators, and processors, all of which require electrical power.

While powered prostheses offer excellent functional restoration, they inevitably lose power occasionally, causing dangerous malfunction. One way to address this issue is incorporation of energy harvesting. The forces and motions of prostheses can be harnessed to mechanoelectrical generators, as depicted in Fig. 9.7. This structure is an air-filled cylinder with a piston attached to a spring-damper actuator. *For example, this could be installed in a leg prosthesis, as shown in* Fig. 9.7.

For example, the pneumatic (air) spring cylinder has radius r, length L, displacement $x(t)$, initial (atmospheric) pressure P_0 and gauge pressure $P(x)$, on top of a mechanical spring with stiffness k and damping c. During walking, the system is subjected to the periodic ground reaction force $F(t)$, which responds as a spring restoring force $F_s(t)$.

The pneumatic air spring restoring force term is given by Eq. (9.1).

$$F_s(t) = P_0 \pi r^2 \left[\left(\frac{L}{L - x(t)} \right)^\gamma - 1 \right]$$

9.9 Walking Energy Harvesting

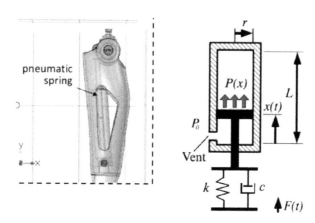

Fig. 9.7 Pneumatic energy harvesting in a prosthetic knee

where F_s is the restoring force of the pneumatic spring mechanism, r is the radius of the pneumatic cylinder, L is the pneumatic cylinder length, x is the piston displacement, P_0 is the initial pressure and is equal to 101,325 Pa at the atmospheric pressure, and γ is the diabatic constant that is equal to 1.4 for diatomic gases such as nitrogen and oxygen.

Alternates to the pneumatic cylinder is piezoelectric material, such as the common PVDF. Typical transduction constants range around 15 picoCoulombs/N.

Mechanical vibration-based energy harvesting systems using piezoelectricity hold the potential for numerous clinical and manufacturer applications. Bone tissue inherently functions as a piezoelectric material operating the ability to generate electricity when subjected to external load. This piezoelectric generation of bone tissue has been shown to enhance growth and proliferation of the tissues themselves, increasing the overall structural rigidity of the area under the mechanical load. Furthermore, osteoblasts and osteoclasts demonstrate increased growth and proliferation when subjected to external piezoelectricity which can give rise to clinical development of piezoelectric nanogenerator devices to expedite bone remodeling processes post fracture.

Piezoelectricity defines the ability of specific materials to generate electrical fields when subjected to mechanical deformation. The mechanoelectrical process of piezoelectricity arises from the unique inversely oriented crystalline structure of piezoelectric materials and neutral electrical charge when in an undeformed state. When the inversely oriented molecules are subjected to deformation, electrical fields form due to the upset of positively and negatively charged poles causing net electrical charges.

Piezoelectric (PZ) devices can transform the energy from force into an electric current, which can be used to power prosthetic motors and controls. (A Numerical Feasibility Study of Kinetic Energy Harvesting from Lower Limb Prosthetics. Yu Jia, Xueyong Wei, Jie Pu, Pengheng Xie, Tao Wen, Congsi Wang, Peiyuan Lian,

Song Xue, and Yu Shi. Energies 2019, 12, 3824; https://doi.org/10.3390/en12203824).

"Oxygen consumption (VO_2) of individual muscles during locomotion using the Fick method" JEB.

9.10 Comparative Energetics

To gain some insight into the role of energy conservation in human walking, we will compare two locomotory creatures whose energy usage is drastically different: the turtle and the penguin. The Galapagos turtle is a slow moving creature that generates minimal inertia, as it pauses briefly at almost every step. Its gait is not rhythmic, but rather lumbering. So how much K.E. is gained from its P.E.? A gait analysis of the turtles was done while estimating their energy usage through measurement of oxygen consumption as depicted in Fig. 9.8.

The P.E. variations at each step of the turtle are about 0.04 J/stride. Its K.E. only fluctuates about ¼ of the P.E., because these energy fluctuations were seldom out of phase. It can be concluded that the efficiency of energy return is about 25% in the turtle. This value is relatively low, and so we can ask, does the turtle expend relatively more energy than other animals in locomotion? Surprisingly, the answer is no; their minimum cost of transport for level walking is 8.0 J/kg/m, as shown in the figure below. Note that the average cost of transport for animals of similar size is 16 J/kg/m. Thus despite having poor mechanical-energy conservation, turtles are very economical at level walking (half the expected metabolic cost). It appears then that the turtle can move without much energy cost, relative to other animals. The most obvious explanation for this efficiency seems to relate to its slowness, which allows for highly efficient use of muscles.

Fig. 9.8 A turtle wearing a gas exchange mask for measuring energy expenditure. Peter A. Zani, Rodger Kram, Audrone R. Biknevicius, and Stephen M. Reilly. THE ENERGETICS AND BIOMECHANICS OF TURTLE LOCOMOTION. ISB XXth Congress—ASB 29th Annual Meeting

9.10 Comparative Energetics

Next consider the penguin, which consumes twice as much energy as other animals their size. The emperor penguin consumes 8.6 J/kg/m for walking, more than that of the turtle and more than twice that of an ostrich (Rhea) of similar weight. Why is it so inefficient? Analysis of energy return does not explain it, because as seen below, there is a nice antiphase relationship between P.E. and K.E. of the C.O.G. In fact the energy return fraction is 80%. The explanation actually is related to muscle usage and is exactly opposite of the case for the turtles. Due to their short legs, penguins must take short, rapid steps to locomote, and hence they use the fastest, least efficient muscles.

Kinetic analyses of turtles were done while estimating their energy usage in terms of oxygen consumption as depicted in Fig. 9.8. It was found that each step costs about 0.04 J/stride, and COT was 8 J/kg/m. It was also found that only 25% of the P.E. gained during steps was returned as K.E. Nevertheless, the turtle gait is relatively energy efficient, since the average COT for animals of the same size is 16 J/kg/m. The possible explanation for this difference is the fact that turtles travel extremely slowly and can use highly efficient muscles, with very low metabolic rates.

Next, we consider the high end of gait activities, such as Olympic running. Several high performance running-specific (RSP) prostheses have been developed for competitive amputee runners, as shown in Fig. 9.9. These epitomize the concept of energy storage and return.

Performance of RSPs depends largely on their stiffness, as measured by β. Stiffness is optimized for particular runners, and its value is a compromise between efficiency in energy returns, and comfort and safety for the runner.

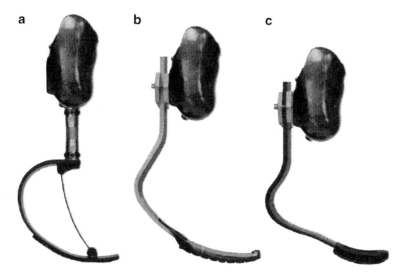

Fig. 9.9 Three running-specific prostheses (**a**) Freedom Innovations Catapult FX6 (C-shaped) configured 2 cm taller than the International Paralympic Committee maximum allowable height (IPC max), (**b**) Oʻssur Flex-Foot Cheetah Xtend (J-shaped) configured at the IPC max and (**c**) Ottobock 1E90 Sprinter (J-shaped) configured 2 cm shorter than the IPC max

The International Paralympic Committee has issued guidelines for these prostheses, and manufacturers vary in their adherence to these guidelines.

Optimizing the design of an RSP for individual runners is done using test beds as shown in Fig. 9.10.

Stiffness of RSPs is tested using a unique Instron machine, as shown in Fig. 9.11.

Fig. 9.10 Testing RSPs. (β) is the calculated angle between the longitudinal axis of the RSP (dashed blue line) and the peak resultant GRF vector (solid red arrow). (β) is a measure of the RSP stiffness

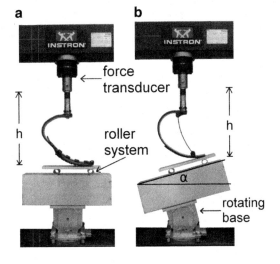

Fig. 9.11 Test setup for assessing RSP stiffness. The manufacturer's rubber sole was attached to the RSPs. Testing was done at various slopes

9.11 Testing with Amputees

Amputee runners running on various RSPs at 3 m/s, experienced a mean peak resultant GRF 2.5 ± 0.3 times body weight. At 6 m/s the average peak resultant GRF was 2.7 ± 0.3 times body weight.

9.12 Problems

1. State a reason that flying should require lower energy than walking.
2. Describe the concept of energy conservation during gait. Use the pendulum or similar device to explain the idea of antiphase energies.
3. Does reducing speed of walking necessarily reduce energy cost? Compare human versus turtle walking in your answer.
4. Prove, using a numerical example (with simulink or other program), that penguin lateral K.E. variation < Fore-aft K.E. variation. (Refer to Penguin.pdf).
5. State how selective action of muscle groups during gait minimizes energy expenditure, and use a specific example.
6. As cadence increases, what happens to other gait parameters?

Chapter 10
Advancing Prosthetic Designs

10.1 Trends in Prosthetic Technology

While prosthetic hardware development has advanced rapidly, the means for humans to control it advances slowly. An example of this trend is evidenced by a 2005 initiative from DARPA, labeled, *Revolutionizing Upper Limb Prosthetics,* which convened a symposium of experts, of which the author was one. Resulting from this initiative was a $100 million investment in UL prosthetic development and several new designs, such as a prosthesis named DEKA. DEKA responds to a combination of user commands, including foot movements, myoelectric signals from the residual limb, and pressure switches located on the body. The DEKA hand, with a power wrist employing two DOFs, is a vast improvement over the current split hook offering a single degree-of-freedom. In addition to the DEKA arm, another device, the *Modular Limb Prosthesis* (MLP), was introduced through the DARPA initiative. The MLP is an advanced prosthesis to be controlled directly by brain signals. Neither of these innovative devices, however, have been universally well reviewed by users.

While DEKA and other advanced UL prostheses are technological marvels, tests with amputees have shown that while some perceived less difficulty in tasks compared with their usual prosthesis, they judged its overall dexterity as "equivalent or somewhat diminished with the DEKA Arm than with the existing prosthesis." Another disappointing conclusion of the test study was, "The majority of participants expressed a preference for the controls of their personal prosthesis and controls rather than the iteration of EMG-PR controlled DEKA Arm used in this study." These user comments are not surprising, because studies have shown that upper-limb prosthetic innovations do not necessarily "translate into functional benefits" [1]. Thus, the technical ability to produce highly functional robotic arms, as well as advanced UL prostheses such as "i-Limb," MichaelAngelo [2], and Bebionic [3], has progressed dramatically, but benefited users little. Currently, researchers are exploring ways to improve the user's interaction with his/her device.

10.2 Focusing on Users

10.2.1 Improving Manipulation

- The value of a prosthesis is best judged by its user, and users of standard myoelectric prosthesis have voiced their opinions in surveys. Their primary critique is lack of dexterity in terms of limited functionality for activities of daily living (ADL) such as personal care, preparing food, and operation of electronics, i.e., cell phones. With practice, persons using a below-elbow prosthesis can adapt to the device and achieve a degree of manipulation ability. For those with higher loss, such as above-elbow amputation and shoulder disarticulation, upper-limb function is quite challenging. The main difficulty for UL prosthetic users is manipulation, a feature that requires independent movements of the thumb, index finger, and wrist, similar to that of the natural limb. Current UL replacements do not embody sufficient degrees of freedom to accomplish such manipulative ability.
- Users of UL prostheses, especially those with above-elbow loss, can benefit from hardware that endows a sense of "feel" for objects, to help gauge the weight, size, and shape of objects they need to manipulate. Prostheses can offer this capability by incorporating force sensors (tactors) on the prosthetic fingers that deliver force feedback somewhere on the residuum. This technique is a valuable feature for users who adapt well to the vibratory pulses on the skin, but many users find the vibration annoying, and alternative methods of sensory substitution are needed.

10.2.2 Sensory Feedback for Dexterity

- The loss of an upper limb removes both motor and sensory functions. The most advanced upper-limb replacements can enable one to pick up a cup of coffee, but cannot deliver the natural sensory feedback that is unconsciously used to feel the grip security and the temperature of the object. In the absence of this type of feedback from the many natural sensors of body position, forces, and nature of the material, limb dexterity is compromised, and restoring it with artificial limbs is very difficult. Ironically, the older-technology body-powered prostheses provide better sensory feedback to the user, since the control cable and body harness provide information about the forces being applied. The more advanced prostheses, i.e., the myoelectric prosthesis, lose these features, because the afferent channels either cease to exist or are severely attenuated.
- Lack of sensory feedback is a primary reason for poor performance of UL prosthesis. Human limbs, especially the hands, can detect minute surface features with exquisite sensitivity. Natural sensors on the skin feed back information to the brain, delivering the finest details of the surface, weight, shape, and stiffness

of objects. These sensory receptors that include both tactile sensors and proprioceptors are widely dispersed throughout muscles, joints, and skin. The sensors report information regarding (1) muscle force and length, (2) velocity of joints, (3) tactile sensation, such as touch and pressure, and (4) temperature. Sensations from these biological sensors include both cutaneous and kinaesthetic cues that can discriminate features of the environment and objects. Intact limbs can sense the size, weight, shape, orientation, and surface properties of objects through a process known as *tactile flow*. This is active when the fingertip slides over the surface of an object to feel its texture, and follows its contours to perceive its shape. Many of our senses from various parts of the body combine to spatially integrate the information coming from its sensors. Restoring this capability to amputees via their prosthesis is a formidable task; it requires a sensory system linking the prosthesis to the brain.

- A schematic of a potential feedback system for UL control is shown in Fig. 10.1.
- Restoring tactile sensation to prosthetic users by sensory substitution is a technique that is useful for some users. The sense of "feel" of objects can be restored to some users by transmitting force signals to appropriate locations on their body, in a process known as sensory substitution. Tactile sensors measure the pressure on the fingers or arm, and vibrotactile stimulation is delivered by electrical signals to the user's skin in proportion to applied pressure. With this approach, the user feels a cutaneous representation, or "tingling," sensation when the prosthesis picks up an object, possibly in proportion to its weight or stiffness.

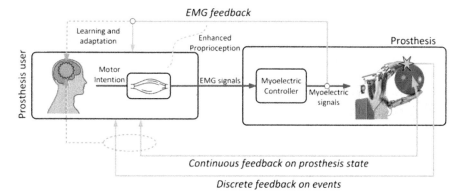

Fig. 10.1 Example of feedback for prosthetic manipulation. This scheme shows a user of a UL prosthesis commanding a grasp of a ball. The motor intention sends EMG signals to the controller which causes the terminal device to grasp the ball. The pressure on the ball and the state of the terminal device send signals back to the user in order to adjust and fine-control the action. The feedback can be direct tactile to the user, or it can be used to adjust the controller directly

Feedback is useful for more than real-time continuous regulation. It can supply information about important events (e.g., contact), enhance natural proprioception (sense of effort), and facilitate learning and adaptation through the development of internal models. Closed loop control of prosthesis. doi: 10.3389/fnins.2020.00345 Jonathon W. Sensinger and Strahinja Dosen Frontiers in Neuroscience | www.frontiersin.org

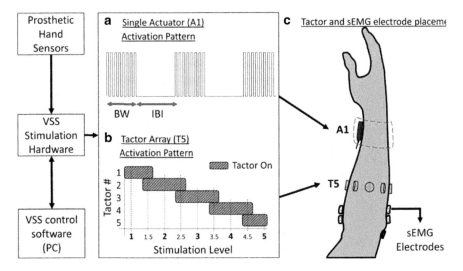

Fig. 10.2 Vibrotactile feedback. Andreas Franzke, Morten B. Kristoffersen, Raoul M. Bongers, Alessio Murgia, Barbara Pobatschnig, Fabian Unglaube, Corry K. van der Sluis PLOS ONE | https://doi.org/10.1371/journal.pone.0220899 August 29, 2019

- A schematic vibrotactile system is shown in Fig. 10.2. Here the vibratory feedback is delivered as bursts of pulses to the transducer A1 on the arm. Activation patterns and amplitudes are variable, in order to transmit information about the degree of pressure on the arm, measured by T5 array. The EMG electrodes also send information about the ongoing muscle activity.
- The strategies described above provide a degree of closed-loop control of the prosthesis, but it is not ideal, since it is considered a nuisance factor by some users, and the current-delivering electrodes can become mechanically unstable and lose function due to natural process of tissue fibrosis. Advanced hands employ biomimetic fingertips made from fluid-filled silicone elastomers to combine compliant grip and tactile sensing. While these features are available, they are limited in function and are far from restoring real feel. Users also report the sensory substitution feels strange and invasive on their skin. User acceptance of such sensory substitutes depends upon whether the benefit in terms of dexterity outweighs the discomfort.
- For lower-limb prostheses, the leg joints can be instrumented with a variety of inertial measurement units, such as accelerometers and load cells in order to report forces to the user, as well as to replace the natural feedback lost in order to control the hardware, as described in Chap. 4.

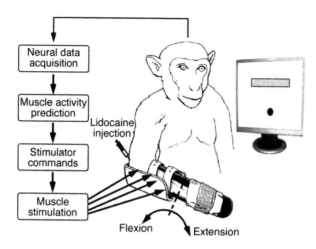

Fig. 10.3 A monkey with arm paralysis is instrumented with electrodes in the motor cortex, which control an arm orthosis. Pohlmeyer EA, Oby ER, Perreault EJ, Solla SA, Kilgore KL, et al. (2009) Toward the Restoration of Hand Use to a Paralyzed Monkey: Brain-Controlled Functional Electrical Stimulation of Forearm Muscles. PLoS ONE 4(6): e5924. doi:10.1371/journal.pone.0005924

10.2.3 Direct Brain Control

The ideal restoration of limb function would be a brain interface directly to the prosthesis, as depicted in Fig. 10.3.

While direct brain control of prostheses is not yet practical for humans, it has been successfully employed to operate an orthosis to restore function of a monkey's paralyzed limb using functional electrical stimulation, in open loop, as shown.

10.3 Engaging Users in UL Prosthetic Design

A segment of 104 potential users of advanced UL prostheses control were surveyed after viewing pictures of the proposed designs, shown in Fig. 10.4. The survey elicited their opinions on the four different control options shown in Fig. 10.4, and results are shown in Table 10.1. The data show that myoelectric control is quite acceptable to 83% of respondents. Also a majority of potential users would be willing to undergo muscle reinnervation and peripheral nerve reinnervation, both of which require surgery and a lengthy recovery period. Only 39% of amputees were interested in a cortical interface.

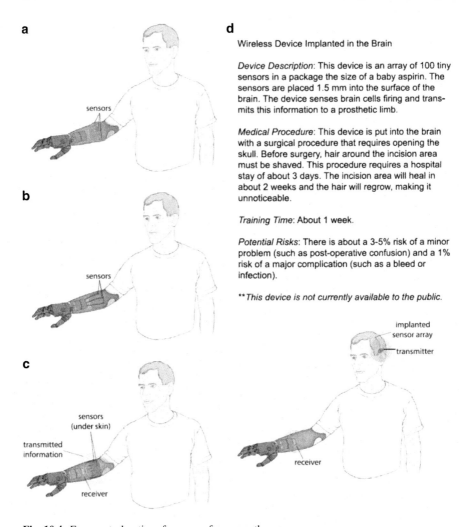

Fig. 10.4 Four control options for users of arm prostheses

Table 10.1 User responses

Interface type	Positive response rate/104 subjects (%)
Myoelectric control	83
Targeted muscle reinnervation	63
Peripheral nerve interfaces	68
Cortical interfaces	39

JOURNAL OF NEUROENGINEERING AND REHABILITATION Volume: 12 Open access Article Number: 53 DOI: 10.1186/s12984-015-0044

10.3.1 Reasons for Abandonment of Prosthesis

Many prosthetic users abandon their prosthesis because it provided little use for their activities of daily living (ADL). One example of such an ADL is the act of eating, which is challenging for users with various types of upper-limb loss operating a myoelectric prostheses. The users were surveyed about their current prosthesis, and then given a new one (not necessarily more advanced) and again asked their experience. Unilateral below-elbow amputees generally thought well of their old prosthesis, but the above-elbow subjects had a considerably lower opinion of their arms, and shoulder disarticulation amputees viewed their prostheses as being of relatively little value. In almost all cases, the amputee rated their new prosthesis more useful than the old in eating, but overall responses indicated a relatively low level of satisfaction. In particular, almost half of the above-elbow users reported that their new prosthesis had "no value" for eating. Each clinical application represents a major engineering achievement, and each one is usually somewhat different from all others. The real limitation in the development of sophisticated upper-extremity systems for the problem of fitting and the nature of disability are so different among the relatively limited numbers of amputee users and congenitally deformed children that the sophisticated engineering required is often economically unjustified. However, the obvious challenge presented by the creation of an artificial human limb continues to fire the imagination of engineers throughout the world, and one may expect continued progress.

Overall results of two surveys are depicted below.

As seen in Fig. 10.5, user ranking of utility depended on their amputation level. Myoelectric control is popular among users with below-elbow (BE) amputations but not with those with above-elbow (AE) amputation. Users were divided accord-

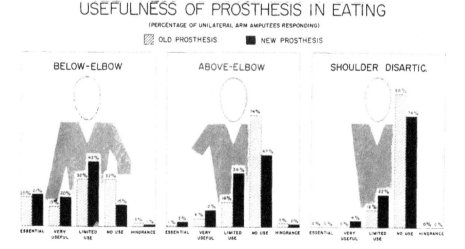

Fig. 10.5 User survey

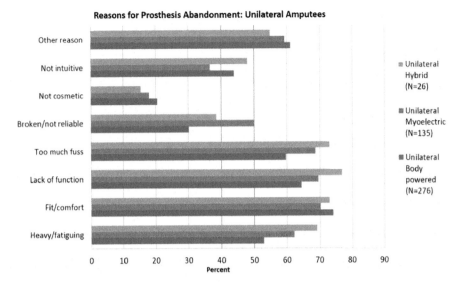

Fig. 10.6 Reasons for abandoning UL prostheses. https://doi.org/10.1371/journal.pone.0213578.g003

ing to the level of amputation, regarding their satisfaction with their prosthesis. Unilateral below-elbow amputees generally thought well of their old prosthesis. Above-elbow subjects, however, have a considerably lower opinion of their prosthesis, and shoulder disarticulation subjects viewed their prostheses of little value. Ease of eating with the prosthesis was a primary concern, and in most cases, subjects rated their new prosthesis more useful than the old in eating. Responses clearly indicate a relatively low level of satisfaction by the users. In particular, almost half of the aboveelbow users reported that their new prosthesis had "no value" for eating. This opinion was characterized primarily by a considerable decrease in the proportion of unilateral amputees (of all types) who considered their prostheses of "no use" or "a hindrance."

For the reasons discussed above, many users abandon their UL prosthesis for various reasons. The primary problem is due to difficulty in control. As seen in the results in Fig. 10.6, nearly 70% of these users found their prosthesis "too much fuss," most probably related to deficiency in the HMI.

10.4 Optimal Restoration of Hand Function

Trans-radial amputations usually leave many arm muscles intact and operational within the residuum. These include the extrinsic hand muscles that control fingers and wrist. Since the control signals for manipulation originate in the brain, these same signals remain operational and capable of fine control of an artificial hand. In

10.4 Optimal Restoration of Hand Function

order to exploit these signals, the user needs to re-train and harness these residual muscles with an efficient HMI, as described in Chap. 9. A variety of training protocols are available, including the SHAP protocol, comprising a battery of practical tests with various objects to both evaluate and train users with their ULP [4]. This test scores based on the time to accomplish certain ADLs.

Modern residual muscle control trainers employ three-dimensional virtual reality (VR) using a realistic prosthesis, as introduced in Chap. 6. The Virtual Interface Adapter (VIA, Fig. 10.7), trains amputee users to control a virtual arm, prior to installation of a prosthesis [5–7]. VIA employs the standard 9-hole peg test of manual dexterity shown in Fig. 10.7. Other tests and virtual games are also available. Before the test, subjects are fitted with a silicone sleeve instrumented with pressure sensors, densely distributed over muscular areas. Users first train the controller to recognize their grasping volition. The subject then places his/her residuum in an instrumented manipulandum, which transmits elbow reaching movements to the virtual environment. The task consists of reaching, grasping, and placing each of the nine pegs, as shown. Amputee subjects, both BE and AE, learned to perform the test successfully after a few minutes training. One subject subsequently practiced a bimanual task on the piano, filmed live by the BBC, as shown in Fig. 10.7 (www.dextrahand.org).

Today, more versatile programs are available for prosthetic simulation training, in particular, MSMS, developed by Loeb & Associates and used by many other labs [2, 8–13]. In addition to training, the MSMS animation tools can be used to alleviate amputee phantom limb pain, as done at Walter Reed National Military Medical Center, where Zeher et al. [14] developed a virtual version of "Mirror Therapy."

The efficacy of training with a prosthetic arm that is either represented virtually, as shown above, or with a robotic arm, is well established [6, 15, 16]. Simulation can also improve grip force control, as shown in a study of subjects with stroke [17], as well as studies of amputees [8].

Customized Training to Use Motor Prosthetic Systems is available with the VR programs available on the MSMS Software, which can simulate various prostheses and allow users to practice on them, as illustrated in Fig. 10.8 [10].

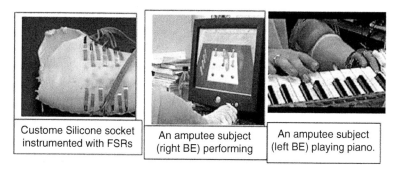

Fig. 10.7 Training with VIA and a piano

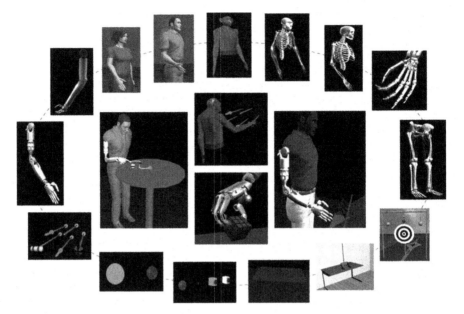

Fig. 10.8 Simulated movements of various prosthetic devices. [10]

10.4.1 Agency as a Factor in HMI Performance

While the HMI is generally the limiting factor for any human's interaction with machines, it is particularly troublesome for amputees, who actually wear the machine in place of their biological limb. Adaptation is not easy because artificial replacements do not match well with users' natural expectations. The perceptual conflict between body image and reality may interfere with the sensorimotor control needed for prosthetic control. The origin of this problem is associated with the user's inability to feel a sense of "agency" with his/her prosthesis. Agency is a recognized factor in users' overall satisfaction and performance with their prostheses [18]. The inability to "own" or fully embody the prosthesis, which may not be consciously recognized, can hinder performance with it. Thus cognitive dissonance within the brain may interfere with volitional control as well as cause phantom pain [15, 19].

When amputees see an illusory limb, there is real-time plastic change in the insula region of their brain, which is a manifestation of agency: the feeling that a person is in control of his or her own body parts (Chambon, Sidarus et al. 2014; de Vries, Byrne et al. 2015; Koban, Corradi-Dell'Acqua et al. 2013). A positive sense of agency contributes positively to the control of limbs, whether they be original or artificial; a negative sense of agency ("error agency") represents a mismatch between reality and expectation, and this interferes with limb control. Error agency is associated with regional brain activations, notably the anterior insula (van Veen, Krug

10.4 Optimal Restoration of Hand Function

et al. 2009). The insula's role in integrating bodily information with sensory inputs with high-level efferent signals coming from the prefrontal and cingulate cortex is well known (Sethi, Suzuki et al. 2012; Allen, Fardo et al. 2016). The nature and possible neural origins of phantom limb (PL) pain are associated with brain regions that respond to incongruence between volition and perceived action [8, 20]. In addition to noxious pain, some amputees experience "referred phantom sensations." The nature and possible neural origins of PL pain have been intensively researched [21] and are associated with brain regions that respond to incongruence between volition and perceived action [8, 20].

Users naturally want a cosmetically correct hand, rather than a hook, but this feature comes at a cost to function, since the gloves are quite stiff and require more effort to operate on the part of the user or on the part of the motor, both of which are subject to fatigue. A more compliant and durable glove material would be an important advance. For this reason, many users are satisfied with just the basic features such as opening and closing the hand slowly, and find the advanced features like fine manipulation and touch sensation to be too complicated. Fortunately, bilateral amputations are relatively rare, and the sound limb can perform these fine functions.

In addition to noxious pain, some amputees experience "referred phantom sensations," wherein stimuli to their prosthetic hand causes homologous sensations in the phantom hand [22]. The *anterior insula* in particular is a brain region that activates during motor-sensory conflicts, which are associated with error agency, i.e., a loss of body ownership or "selfness" [23–25]. Insula activity has also been associated with pain [26]. While error agency is reflected in the insula, a positive sense of agency correlates with activity in the parietal and prefrontal cortical areas as measured by a rise in oxyhemoglobin levels, in response to a real-time virtual tool experiment [27]. Further evidence of parietal cortex involvement with agency is provided by a study on amputees in which the size of their posterior parietal cortex correlated inversely with their amount of prosthetic usage and also with the intensity of PL pain they experienced [5]. These results underscore the role of agency in PL sensations and its localization to particular brain regions.

Successful strategies to mitigate negative and promote positive agency include mirror imagery [9, 28–30], and sensory feedback from surrogate and virtual limbs [10, 11, 31–33]. A contrasting result was reported wherein PL pain intensified when amputee subjects used their brain signals to control an advanced prosthetic hand; surprisingly, pain was reduced when subjects learned to dissociate their use of a prosthetic hand from their PL [34]. Thus, manipulating perception can profoundly influence clients' neural activity, which may or may not be desirable.

Brain scanning with fMRI provides insight on the role of agency in persons both with and without intact upper limbs. One study examined brain activities of subjects with intact limbs while they performed repetitive grasping by their right hand in three conditions: (1) eyes closed, (2) watching their moving hand, and (3) watching a virtual prosthetic hand which they controlled with myoelectrical signals from their arm [17]. Results showed that activity in the right posterior parietal cortex was seen in both conditions 1 and 2, but condition 3 shifted the activity laterally from that of condition 1 and also produced unique activity in the right ventral premotor cortex.

These patterns may represent a perceptual assimilation of the virtual hand into the body schema. Another study of amputees who simply viewed a moving virtual limb demonstrated activations in their contralateral (appropriate) motor cortex [6].

10.5 Restoring Agency

Agency, or the feeling of body ownership, is a well-recognized factor in users' overall satisfaction and performance with their prostheses ([35]; Crawford 2015). Our previous study on brain changes during virtual manipulation by amputees has shown that superimposing a mirror image arm on amputees induces immediate changes in the fMRI patterns in the brain, suggesting that motor learning may be facilitated by visual appearance of a realistic limb [16]; similar studies have supported this notion [36]. There may be simpler methods than fMRI by which to categorize users' responses to prosthetic restoration, such as near-infrared spectroscopy (NIRS).

10.6 Developing Better HMIs

To explore better options for prosthetic control, a scientific panel, the Peripheral Nervous System-Machine Interface (PNS-MI) group, was established [23]. This group identified potentially more robust and accessible sources of control by exploiting the mechanical dynamics of residual muscle, both agonist and antagonist, which encode volition more directly than does the EMG. For example, introduced in Chap. 6, surface muscle pressure (also called, FMG) expresses the entire volitional signal, including trajectory and force, in a highly processed and invariant manner [28, 37]. For users who retain some residual arm muscles, FMG accurately codes for the motion of several DOFs, as well as force, and is superior to EMG in terms of resolution and practicality. The FMG controller, *Dextra*, was used by a person with trans-radial amputation playing a short piano piece in a BBC special and other media after only a few minutes training. Several other HMI options are available to try, including ultrasound [38], optics [39], and NIRS [19].

10.6.1 Hardware Needs

Available UL hardware ranges from the simple prehensile hook having 1 DoF to hands with up to 22 DoFs, multiple axes wrists, as well as powered shoulders and elbows [21]. Advances in technology have been promoted by governmental, commercial, nonprofit, and academic communities initiatives. While some of these new prosthetic components represent true marvels of engineering, they do not always improve user function, as discussed at the beginning of this chapter Specifically, actuators are not ready for the task due to their weight and large power requirements that cannot be met with the presently available power storage components, i.e., batteries.

10.7 Exploiting Adaptability

The human neuromotor system rapidly learns and adapts to changing tasks and environments. Adaptation to foreign appendages, such as prosthetic limbs, therefore, should be achievable, given the appropriate equipment. Current prostheses necessarily use some pre-programmed movements, because of the inherent limitations of the human interface. This feature is useful, but given the almost limitless variety of manual motions humans make is so variable (probably infinitely) that no library of pre-programmed motions could satisfy a user for any length of time. The concept of programming the trajectory of the terminal device so as to limit the decision-making demand upon the user to commanding the system to move it from A to B is thus questionable.

Ideally, the prosthesis should link directly with the central nervous system and go "online" with the prosthetic control system. Myoelectric control is a beginning; however, the available command signal taps the middle of the volition, not its beginning (which originates in the brain), of the efferent loop. Sensory information feeds back information from joints by afferent channels that current prostheses do not receive. Thus information about the output of the HMI can only be inferred rather than known. A more intuitive control system could possibly be based on position-controlled servos, or surface muscle pressure. Agency, as indicated by absence of insula activity in the brain, is immediately restored to the amputee subjects by presentation of an illusory restoration of their hand. Since insula activity is associated with phantom limb pain [40] and may interfere with motor control circuits in the brain [26], its removal can be a useful target for reducing pain and improving the human-machine interface (Maruishi, Tanaka et al. 2004; [41]). These and other studies give strong evidence that user control and comfort with a prosthesis are critically dependent on his/her feeling of ownership of the device.

10.8 Agency as a Key to Controlling the Prosthesis

As noted above, a critical but generally overlooked factor in fuller restoration of function to amputees is agency between user and prosthesis. There is ample evidence, noted above, that training which directly addresses this issue is likely to improve prosthetic restoration of function.

One user characteristic that likely influences agency but has received minimal attention is arm laterality [42, 43]. It is well known that the limbs on either side specialize for different functions, and this relates to cerebral hemispheric differences

[44–47]. In general for the UL, the nondominant limb specializes in positioning and holding objects, while the dominant limb specializes in operations, usually ballistic, on the object held by the nondominant limb. With regard to the lower limb, the two legs may play different roles during walking, such as leading versus following. These differences could impact outcomes and are important considerations in prosthetic prescriptions.

10.8.1 Matching Prosthesis to User

One strategy to optimize restoration is to better systematize the prescription of final prosthetic installation. Since it would be cumbersome to test a large array of components on potential users, simulations could streamline the procedure. Using a sequence of Simulate, Test and Practice (STP), the prosthetist could test many variables and permutations of designs in a systematic, quantitative manner, with minimal work by the client. Physically testing the growing array of prosthetic options, such as the EAD components and HMIs shown in Figs. 10.4, 10.5, 10.6, would require clinics to stock a large, up-to-date inventory and assemble complex hardware for each trial. Simulated hardware and environments might be a better approach. The STP system would also test capabilities of the users residua as they practiced on virtual hardware.

Virtual prosthetic equipment and environments, developed at the Rutgers lab and elsewhere, as described above, have demonstrated their utility in training prosthetic usage, but none have reached the "plug and play" stage necessary for clinical usage. By developing STP in close collaboration with the clinics and patients, the system could be targeted for optimizing outcomes for each patient and convenience for the clinician. Here, we plan to deliver a solid platform for systematizing the prosthetic prescription process, tested on a small number of users, and readied for a formal clinical test.

The STP method is an adaptation of a strategy called implicit motor control (IMC), developed by Dr. Artemiadis and colleagues for UL restoration [15]. IMC requires neither training sets nor correlation with natural hand-arm kinematics. Instead, available signals from residual muscles are remapped to robustly and simultaneously control a multiple degree of freedom prosthetic device. Users can learn arbitrary muscle mappings similarly to learning new motor skills. Mappings learned during one set of tasks naturally transfer to enhanced initial performance when using the same mapping to perform other tasks. New muscle synergies are likely to develop from this type of training.

The STP paradigm would systematically test various combinations of HMI and EAD in a virtual prosthetic arm. For IPPs, the HMI-I will use standard means, including bi-scapular abduction and dual-control arm flexions/extensions, registered by mechanical sensors in the body harness. For powered control, various HMI-E options of Fig. 10.4 will be controlled by a dense array of sensors, either standard or customized for each subject (Fig. 10.7). The HMI-E sensor array will

10.8 Agency as a Key to Controlling the Prosthesis

include MYO, SMP, and other appropriate sensors of neuromuscular activity; sensors will be distributed over the relevant surfaces of the sleeve. For unilateral amputees, and for sound-body subjects, the sound arm(s) will interact with the virtual environment via an instrumented arm sleeve(s) and glove(s).

The overall STP scheme is depicted in Fig. 10.9. A library of components and systems will be developed from manufacturers' specifications and published studies. These will be replicated to the degree possible, using CAD modeling of hardware, and software algorithms. Control software for simulating HMI algorithms will be programed *de novo* or adapted from previous programs, such as *Dextra* and *BioPatRec*. The user will familiarize with the virtual environment by bimanual "playing" with objects, with feedback from simulation. Users will then perform specified tasks on a sequence of virtual prostheses and environments, determined by the prosthetist in conjunction with stated needs and desires of the user. Quantitative performance measures will then establish a short list of potentially optimal prostheses, from which an optimum from available products will be determined.

Prescriptions of prostheses are based on results of iterative evaluations of various options, which include self-reporting and unbiased performance metrics, if available. Examples of these include SHAP (Light CM 2002), performance metrics of advanced prostheses [48], and measures of motor control for position, velocity, and force, to be measured with inertial and imaging sensors [17, 30]. An optimal prosthetic combination (OPC) usually includes commercially available components and open external assistive devices (EADs, i.e., elbow, wrist, and hand), with a specific HMI. The iterative selection process is depicted in Fig. 10.10.

Restoration Outcome metrics: For each client, a unique set of outcome measures should be identified according to that individual's needs profile. For each input combination tested, predictive models of each set of outcome metrics will be generated using standard feature selection processes. We will take a heavily reductionist approach, selecting a subset of metrics, so determined in order to limit complexity in subsequent analysis stages.

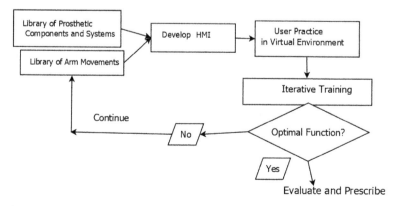

Fig. 10.9 Prosthetic Practice and Prescription

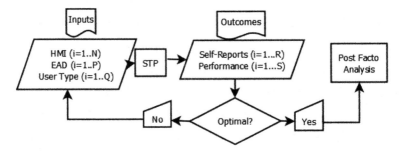

Fig. 10.10 Summary of Simulate-Test-Practice Method

Treatment combination optimization: Following identification of key outcome features, novel treatment combinations can be simulated as described above; these simulated devices, in combination with the subject's unique characteristics, and his/her unique and specific task demands, will yield a new feature space which will further inform the spectrum of outcomes for that subject. The search space will be unique to each subject and will reflect anatomical and personal preferences including handedness, tolerability, cosmetic appeal, and appropriateness. If the search space is adequately small, an exhaustive search will yield the OPC; if the search space becomes intractably large, the OPC will be identified through a Monte Carlo approach.

10.9 Conclusion

The foregoing treatise of some of the historical advances in artificial restoration of limbs and the potential upcoming innovations illustrates the power of incremental advancements made both from within and outside of the specialized field of prosthetics. At a broader view, this pattern is common to the rapidly developing field of bionics.

10.10 Exercises

1. Show control flowcharts of a manually operated UL prosthesis and a motorized prosthesis. Label all operations.
2. Find an example of the role of agency in prosthetic control and summarize it briefly.

Bibliography

1. Resnik L, Borgia M. User ratings of prosthetic usability and satisfaction in VA study to optimize DEKA arm. J Rehabil Res Dev. 2014;51(1):15–26.
2. Soares A, Andrade A, Lamounier E, Carrijo R. The development of a virtual myoelectric prosthesis controlled by an EMG pattern recognition system based on neural networks. J Intell Inform Syst. 2003;21(2):127–41.
3. Steeper R. Bebionic hand. 2014. www.bebionic.com.
4. Light CM, Chappell PH, Kyberd PJ. Establishing a standardized clinical assessment tool of pathologic and prosthetic hand function: normative data, reliability, and validity. Arch Phys Med Rehabil. 2002;83(6):776–83.
5. Craelius W. BBC television show on Dextra. Front Med. 2001.
6. Kuttiva M, Burdea G, Flint J, Craelius W. Manipulation practice for upper-limb amputees using virtual reality. Presence. 2005;14(2):175–82.
7. Kuttiva M, Flint J, Burdea G, Craelius W. VIA: a virtual interface for the arm of upper-limb amputees. In: Second international workshop on virtual rehabilitation: IWVR, Piscataway; 2003.
8. Bouwsema H, van der Sluis CK, Bongers RM. Effect of feedback during virtual training of grip force control with a myoelectric prosthesis. PLoS One. 2014;9(5):e98301.
9. Davoodi R, Loeb GE. Development of a physics-based target shooting game to train amputee users of multijoint upper limb prostheses. Presence. 2012;21(1):85–95.
10. Davoodi R, Loeb GE. Real-time animation software for customized training to use motor prosthetic systems. IEEE Trans Neural Syst Rehabil Eng. 2012;20(2):134–42.
11. Hauschild M, Davoodi R, Loeb GE. A virtual reality environment for designing and fitting neural prosthetic limbs. IEEE Trans Neural Syst Rehabil Eng. 2007;15(1):9–15.
12. Manal K. Real-time control of an EMG-driven virtual arm. Med Sci Sports Exerc. 2004;36(5):S1.
13. Resnik L, Etter K, Klinger SL, Kambe C. Using virtual reality environment to facilitate training with advanced upper-limb prosthesis. J Rehabil Res Dev. 2011;48(6):707–18.
14. Zeher MJ, Armiger RS, Burck JM, Moran C, Kiely JB, Weeks SR, Tsao JW, Pasquina PF, Davoodi R, Loeb G. Using a virtual integration environment in treating phantom limb pain. Stud Health Technol Inform. 2011;163:730–6.
15. Antuvan CW, Ison M, Artemiadis P. Embedded human control of robots using myoelectric interfaces. IEEE Trans Neural Syst Rehabil Eng. 2014;22(4):820–7.
16. Li K, Beetel R, Craelius W. Seeing an imaginary arm alters brain functions of amputees. In: HCI international: 17th international conference on human-computer interaction, Los Angeles; 2015.
17. Kim NH, Wininger M, Craelius W. Training grip control with a Fitts' paradigm: a pilot study in chronic stroke. J Hand Ther. 2010;23(1):63–71; quiz 72.
18. Resnik L, Latlief G, Klinger SL, Sasson N, Walters LS. Do users want to receive a DEKA Arm and why? Overall findings from the Veterans Affairs Study to optimize the DEKA Arm. Prosthet Orthot Int. 2013;38:456–66.
19. Batula AM, Mark JA, Kim YE, Ayaz H. Comparison of brain activation during motor imagery and motor movement using fNIRS. Comput Intell Neurosci. 2017;2017:5491296.
20. Boostani R, Moradi MH. Evaluation of the forearm EMG signal features for the control of a prosthetic hand. Physiol Meas. 2003;24(2):309–19.
21. Belter JT, Segil JL, Dollar AM, Weir RF. Mechanical design and performance specifications of anthropomorphic prosthetic hands: a review. J Rehabil Res Dev. 2013;50(5):599–617.
22. Bridges MM, Para MP, Mashner MJ. Control system architecture for the modular prosthetic limb. J Hopkins APL Tech Dig. 2011;30(3):217–22.
23. Castellini C, Artemiadis P. PNS-MI community. 2013. http://pnsinterfaces.wordpress.com/.
24. Castellini C, Artemiadis P, Wininger M, Ajoudani A, Alimusaj M, Bicchi A, Caputo B, Craelius W, Dosen S, Englehart K, Farina D, Gijsberts A, Godfrey SB, Hargrove L, Ison M, Kuiken T,

Markovic M, Pilarski PM, Rupp R, Scheme E. Proceedings of the first workshop on peripheral machine interfaces: going beyond traditional surface electromyography. Front Neurorobot. 2014;8:22.
25. Chadwell A, Kenney L, Thies S, Galpin A, Head J. The reality of myoelectric prostheses: understanding what makes these devices difficult for some users to control. Front Neurorobot. 2016;10:7.
26. Chand GB, Dhamala M. Interactions between the anterior cingulate-insula network and the fronto-parietal network during perceptual decision-making. Neuroimage. 2017;152:381–9.
27. Connolly C. Prosthetic hands from touch bionics. Ind Robot. 2008;35(4):290–3.
28. Craelius W. The bionic man: restoring mobility. Science. 2002;295(5557):1018–21.
29. Craelius W. A revolutionary upper-limb prosthesis. In: DARPA revolutionizing prosthetics workshop, Arlington; 2005.
30. Craig J, Korenczuk C, Kosinsky C, Newby N, Craelius W, Ma S, Escaldi S. Measuring joint ROM and stiffness with a cell phone. In: Association of Academic Physiatrists annual meeting, San Antonio; 2015.
31. Dorrance D. Artificial hand. United States; 1912.
32. Farina D, Jiang N, Rehbaum H, Holobar A, Graimann B, Dietl H, Aszmann OC. The extraction of neural information from the surface EMG for the control of upper-limb prostheses: emerging avenues and challenges. IEEE Trans Neural Syst Rehabil Eng. 2014;22(4):797–809.
33. Hanger. Michael Angelo Hand. 2013. http://www.hanger.com/prosthetics/services/Technology/Pages/MichelangeloHand.aspx.
34. Ison M, Artemiadis P. Enhancing practical multifunctional myoelectric applications through implicit motor control training systems. In: 36th annual international conference of the IEEE Engineering in Medicine and Biology Society (EMBC), Chicago; 2014.
35. Pirowska A, Wloch T, Nowobilski R, Plaszewski M, Hocini A, Menager D. Phantom phenomena and body scheme after limb amputation: a literature review. Neurol Neurochir Pol. 2014;48(1):52–9.
36. Yokoi H, Arieta AH, Katoh R, Yu WW, Watanabe I, Maruishi M. Mutual adaptation in a prosthetics application. Embodied Artif Intell. 2004;3139:146–59.
37. Abboudi RL, Glass CA, Newby NA, Flint JA, Craelius W. A biomimetic controller for a multifinger prosthesis. IEEE Trans Rehabil Eng. 1999;7(2):121–9.
38. Akhlaghi N, Baker CA, Lahlou M, Zafar H, Murthy KG, Rangwala HS, Kosecka J, Joiner WM, Pancrazio JJ, Sikdar S. Real-time classification of hand motions using ultrasound imaging of forearm muscles. IEEE Trans Biomed Eng. 2016;63(8):1687–98.
39. Nissler C, Mouriki N, Castellini C. Optical myography: detecting finger movements by looking at the forearm. Front Neurorobot. 2016;10:3.
40. Willoch F, Rosen G, Tolle TR, Oye I, Wester HJ, Berner N, Schwaiger M, Bartenstein P. Phantom limb pain in the human brain: unraveling neural circuitries of phantom limb sensations using positron emission tomography. Ann Neurol. 2000;48(6):842–9.
41. Powell MA, Kaliki RR, Thakor NV. User training for pattern recognition-based myoelectric prostheses: improving phantom limb movement consistency and distinguishability. IEEE Trans Neural Syst Rehabil Eng. 2014;22(3):522–32.
42. Romkema S, Bongers RM, van der Sluis CK. Intermanual transfer in training with an upper-limb myoelectric prosthesis simulator: a mechanistic, randomized, pretest-posttest study. Phys Ther. 2013;93(1):22–31.
43. Wang JS, Sainburg RL. The dominant and nondominant arms are specialized for stabilizing different features of task performance. Exp Brain Res. 2007;178(4):565–70.
44. Morris T, Newby NA, Wininger M, Craelius W. Inter-limb transfer of learned ankle movements. Exp Brain Res. 2008;192(1):33–42.
45. Sainburg RL. Evidence for a dynamic-dominance hypothesis of handedness. Exp Brain Res. 2002;142(2):241–58.
46. Sainburg RL, Kalakanis D. Differences in control of limb dynamics during dominant and nondominant arm reaching. J Neurophysiol. 2000;83(5):2661–75.

47. Zhang W, Sainburg RL, Zatsiorsky VM, Latash ML. Hand dominance and multi-finger synergies. Neurosci Lett. 2006;409(3):200–4.
48. Resnik L, Borgia M, Latlief G, Sasson N, Smurr-Walters L. Self-reported and performance-based outcomes using DEKA Arm. J Rehabil Res Dev. 2014;51(3):351–62.
49. Alphonso AL, et al. Use of a virtual integrated environment in prosthetic limb development and phantom limb pain. In: Wiederhold BK, Riva G, editors. Annual review of cybertherapy and telemedicine 2012: Advanced technologies in the behavioral, social and neurosciences; 2012.
50. Ameri A, Scheme EJ, Kamavuako EN, Englehart KB, Parker PA. Real-time, simultaneous myoelectric control using force and position-based training paradigms. IEEE Trans Biomed Eng. 2014;61(2):279–87.
51. Mattar E. A survey of bio-inspired robotics hands implementation: new directions in dexterous manipulation. Robot Auton Syst. 2013;61(5):517–44.
52. Natarajan GS, Wininger M, Kim NH, Craelius W. Relating biceps EMG to elbow kinematics during self-paced arm flexions. Med Eng Phys. 2011.
53. Newby N. DextraHand. 2011. www.dextrahand.org.
54. Pylatiuk C, Doderlein L. "Bionic" arm prostheses. State of the art in research and development. Orthopade. 2006;35(11):1169.
55. Pylatiuk C, Schulz S, Doderlein L. Results of an Internet survey of myoelectric prosthetic hand users. Prosthet Orthot Int. 2007;31(4):362–70.
56. Resnik L, Klinger S. Attrition and retention in upper limb prosthetics research: experience of the VA home study of the DEKA arm. Disabil Rehabil Assist Technol. 2017;12(8):816–21.
57. Resnik L, Klinger SL, Etter K. The DEKA Arm: its features, functionality, and evolution during the Veterans Affairs Study to optimize the DEKA Arm. Prosthet Orthot Int. 2013.
58. Resnik LJ, Borgia ML, Acluche F. Perceptions of satisfaction, usability and desirability of the DEKA Arm before and after a trial of home use. PLoS One. 2017;12(6):e0178640.
59. Scott TRD, Haugland M. Command and control interfaces for advanced neuroprosthetic applications. Neuromodulation. 2001;4(4):165–74.
60. Smit G, Bongers RM, Van der Sluis CK, Plettenburg DH. Efficiency of voluntary opening hand and hook prosthetic devices: 24 years of development? J Rehabil Res Dev. 2012;49(4):523–34.
61. Weir RF. The great divide - the human-machine interface: issues in the control of prostheses, manipulators, and other human machine systems. In: IEEE 29th annual northeast bioengineering conference; 2003.
62. Weir RF. Design of advanced prosthetic limb systems. In: Castelli VP, Troncossi M, editors. Grasping the future: advances in powered upper limb prosthetics. Sharjah: Bentham Science; 2012. p. 3–14.
63. Mayer RM, Garcia-Rosas R, Mohammadi A, Tan Y, Alici G, Choong P, Oetomo D. Tactile feedback in closed-loop control of myoelectric hand grasping: conveying information of multiple sensors simultaneously via a single feedback channel. Front Neurosci. 2020;14:348. https://doi.org/10.3389/fnins.2020.00348.

Appendix: Computational Modeling

Simulink

This chapter presents tools to build and analyze mathematical models in Simulink and Matlab. Models help us understand and manipulate complex biological processes under a variety of conditions; they allow real-time examination *in silico* of processes that are difficult or impossible to observe *in vivo*. A model represents biomechanical functions unambiguously, enabling diverse researchers to use and test the same system with a clearly defined hypotheses built into the model. This feature is appealing for modeling, more so than empirical data collection, because it allows for replicable experimentation under reproducible conditions, which is crucial to scientific inquiry and progress.

Models can approximate the many diverse individual mechanisms operating within an overall working system, which can be tested in a wide range of variations. This primary feature can accelerate our understanding of biological systems, without having to do biological studies, which have limited testing possibilities. Now, there is a large and growing body of data on human biomechanics that are highly representative and can be embedded mathematically into models. System behavior thus can be simulated and manipulated to make highly accurate predictions in various scenarios. This is especially relevant to prosthetic designs, since models are invaluable tools for testing designs, without risk to the client.

Encapsulating existing biomechanical knowledge in mathematical formulations compresses complex mechanisms into a compact format for testing hypotheses. Models derived from the operational rules of mathematics are inherently self-consistent and an ideal framework for manipulating multiple system variables. Hypotheses that are embedded in a model can be systematically tested by comparing predicted behavior with actual behavior produced by the model.

It is important to acknowledge that models *are simplified representations*, aka simulacra, *that reflect* one's best guess regarding the underlying processes. The power of modeling lies in its replication of the scientific method: the discrepancy between model prediction and physiological observation. More often than not, initial models are incorrect or overly simplistic and may not represent true behavior. The model allows for iterative trials of designs, wherein results are "fed back" to the model for parameter adjustments. Yet, the power of the modeling process lies in its replication of the scientific method: the discrepancy between model prediction and physiological observation is "fed back" to the model, so that parameters can be adjusted. This allows us to return to the model development stage once again in order to modify our previous assumptions. Subsequently, we would retest the revised model against experimental observations, and so the alternating process of induction and deduction continues until we are satisfied that the model can predict the observation, and even "explain" most of the observed behavior. Then, having arrived at a "good" model, we could venture to use this model to predict how the system might behave under experimental conditions that have not been employed previously. These predictions would serve as a guide for the planning and design of future experiments.

Learning from Models

The main purpose of a model is to act as a platform for testing hypotheses on how the musculo-skeletal system works, and how it can be augmented or repaired by artificial means. The first step in designing any device is to establish its specifications: what it needs to accomplish the job. In modeling prostheses, you are attempting to test designs for optimum performance in activities of daily living (ADLs). The main value of a model is being a platform for testing hypotheses on how the musculo-skeletal system works, and how it can be augmented or repaired by artificial means. The first step in designing any device is to establish its specifications: what it needs to accomplish the job. In modeling human motion, you are attempting to re-design anatomical structures in order to replace missing parts of the body and restore optimum performance in ADLs. Many tools are available to help this process, including drawing, physical calculations, measurements, prototyping, and modeling. The latter tool can incorporate results from all available sources and test design integrity.

Even a small amount of data applied to a model can lead to information and design solutions. They usually start with a conceptual idea of physical system whose behavior you would like to visualize. Once you make the model, you can test its output in response to any input.

Mathematical modeling adds detail to the conceptual model. However, unlike verbal languages, mathematics provides descriptions that are unambiguous, concise, and self-consistent. The general purpose of a mathematical model is to calculate quantities that are difficult or impossible to measure. A model enables different researchers to use and test the same system without being confused about the hypotheses built into the model. Since the equations employed in the model are based, at least in large part, on existing knowledge of the physiological processes in question, they also serve the useful purpose of archiving past knowledge and compressing all that information into a compact format. The inherent self-consistency of the model derives from the operational rules of mathematics, which provide a logical accounting system for dealing with the multiple system variables and their interactions with one another. On the other hand, the hypotheses embedded in some components of the model *are* only *hypotheses, reflecting* our best belief regarding the underlying process. Often, these initial ideas are inexact and/or overly simplistic. As a consequence, the behavior of the model may not reflect the corresponding reality. Yet, the power of the modeling process lies in its replication of the scientific method: the discrepancy between model prediction and physiological observation can be used as "feedback" to alert us to the inadequacies of one or more component hypotheses. This allows us to return to the model development stage once again in order to modify our previous assumptions. Model revisions should be retested against experimental observations, and the alternating process of induction and deduction continues until the model produces satisfactory results. Then, having arrived at a "good" model, we venture to use it to predict how the system might behave under experimental conditions that have not been employed previously. These predictions would serve as a guide for the planning and design of future designs.

Many tools are available to help the modeling process, including drawing, physical calculations, measurements, prototyping, and modeling. The latter tool can incorporate results from all the previous processes and test the integrity of the design.

In summary, models help deduce information and can make predictions from a small amount of data. They usually start with a conceptual idea of physical system whose behavior you would like to visualize. Once you make the model, you can test its output in response to any input.

Approaching the Problem

Most real-world problems are ill defined. For example, suppose you are designing a lower-limb prosthesis. You are trying to optimize its weight, balance, length, and functional operation. For modeling, usually the best first step is to sketch or write down all the parameters and variables involved. If you miss some variables at first, that's okay—you can correct them as you proceed. The idea is to start gathering information, i.e., the very basics: what kind of knee would work best for my client? What magnitude of forces would it experience? What kind of activities does the client do? Is the knee appropriate for a K-Level 1 (ambulating skill over level surfaces), or a K-Level 4 (active adult or athlete)? Then, you can assemble more sophisticated descriptions, for example, how much force will the prosthesis be subjected to, and where are the forces being applied? What is the desired range of motion?

The second step is to formalize the problem with a free body diagram (FBD). This is a sketch that reduces all components of the system to the "bare bones," where even skeletal bones can be modeled as sticks. If the structure consists of multiple segments, depending on the nature of the analysis, each segment can be represented as a mass at the center point, the center of mass (COM), and a moment of inertia. If the goal is analyzing the structure as a whole, the segment COMs can be lumped into a single point. Next the external forces and moments acting on each segment are drawn. From this sketch, Newton's laws can be directly applied.

Block Diagrams

Block diagrams can help clarify the relationship among key variables of the system under study. The block diagram therefore represents a *conceptual model* of the system under study. Block models can merely show relationships or flow among components, and they can be used to represent quantitative calculations. Simple math operations, for example, can be represented as connected operational blocks (Figs. A.1 and A.2).

Typically, block diagrams represent a controlled system, using negative feedback, as shown.

> https://graphicdesign.stackexchange.com/questions/16629/drawing-block-diagram

A control model allows us to hypothesize about the contents in each of the "boxes" of the block diagram. Each of the boxes represents a transfer function, expressed as a differential equation. In this case, a process, P, is controlled by negative feedback.

Generic open-loop and closed-loop control systems are shown in Fig. A.3

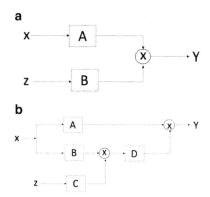

Fig. A.1 (**a**) Block model $Y = Ax + Bz$. (**b**) The equation being solved is $Y = Ax + D(Bx - Cz) = Y$. Any degree of algebraic complexity can be represented

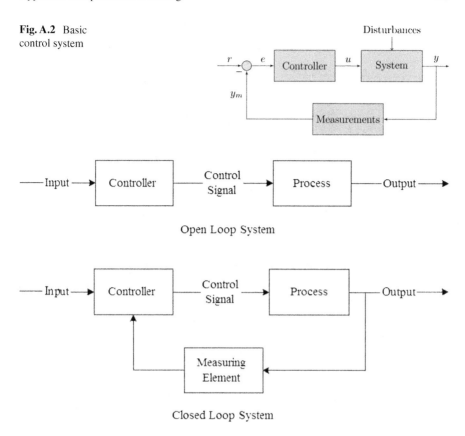

Fig. A.2 Basic control system

Fig. A.3 Open and closed loop control

A practical example of a human feedback control is the model of the knee-jerk reflex, Fig. A.4.

The box labeled "reflex center" in the spinal cord will contain an expression of our belief about how the change in afferent neural frequency may be related to the change in efferent neural frequency. This relationship may be linear or nonlinear. Many questions need to be considered. For example, how does the time-course of the response in efferent frequency follow the changes in afferent frequency? One way of answering these questions would be to isolate this part of the physiological control system and perform experiments that would allow us to measure this relationship.

In this case, the relationship between the controller input and controller output is derived purely on the basis of observations, and therefore, it may take the form of a table or a curve, best fit to the data. Alternatively, these data may already have been measured, and one can use real data from the literature to establish the required input-output relationship. This kind of model assumes no internal structure and has been given a number of labels in the physiological control literature, such as *black-box, empirical,* or *nonparametric* model. Frequently, on the basis of previous

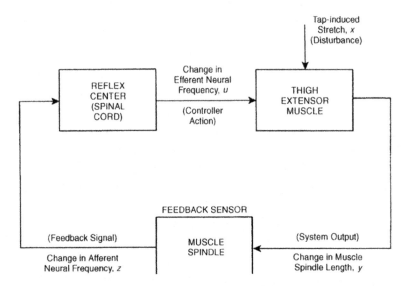

Fig. A.4 Knee jerk reflex

knowledge, we also have some idea of what the underlying physical or chemical processes are likely to be. In such situations, we might propose a hypothesis that reflects this belief. On the basis of the particular physical or chemical laws involved, we would then proceed to derive an algebraic, differential, or integral equation that relates the "input" to the "output" of the system component we are studying. This type of model is said to possess an internal structure, i.e., it places some constraints on how the input may affect the output. As such, we might call this a *structural* or *gray-box* model. In spite of the constraints built into this kind of model, the range of input-output behavior that it is capable of characterizing can still be quite extensive, depending on the number of free parameters (or coefficients) it incorporates. For this reason, this type of model is frequently referred to as a *parametric* model.

Mathematical modeling adds detail to the conceptual model. However, unlike verbal languages, mathematics provides descriptions that are unambiguous, concise, and self-consistent, and contextual (i.e., strictly rule-based). The purpose of a mathematical model is to calculate quantities that are difficult or impossible to measure. A model enables different researchers to use and test the same system without being confused about the hypotheses built into the model. Since the equations employed in the model are based, at least in large part, on existing knowledge of the physiological processes in question, they also serve the useful purpose of archiving past knowledge and compressing all that information into a compact format. The inherent self-consistency of the model derives from the operational rules of mathematics, which provide a logical accounting system for dealing with the multiple system variables and their interactions with one another. On the other hand, the hypotheses embedded in some components of the model *are* only *hypotheses*, *reflecting* our best belief regarding the underlying process. More often than not,

Appendix: Computational Modeling

these are inexact or overly simplistic. As a consequence, the behavior of the model may not reflect the corresponding reality. Yet, the power of the modeling process lies in its replication of the scientific method: the discrepancy between model prediction and physiological observation can be used as "feedback" to alert us to the inadequacies of one or more component hypotheses. The feedback can tell us how to modify our model once again in order to modify our previous assumptions. Subsequently, we would retest the revised model against experimental observations, and so the alternating process of induction and deduction continues until we are satisfied that the model can explain most of the observed behavior. Then, having arrived at a satisfactory model, we could venture to use this model to predict how the system might behave under experimental conditions that have not been employed previously. These predictions would serve as a guide for the planning and design of future experiments.

Models help deduce information and can make predictions from a small amount of data. They usually start with a conceptual idea of physical system whose behavior you would like to visualize. Once you make the model, you can test its output in response to any input.

Sample Models: In this section, the mechanical behavior of select cellular components will be modeled in Simulink. Your main tools are free body diagrams, block diagrams, and Simulink.

Steps for modeling:

– Simplify the anatomy
– Find the right formulae
– Derive equations of motion
– Check by dimensional analysis
– Set up all blocks and operations
– Execute the code
– Check and verify results
– Add constraints to model
– Test the model with different inputs

Model data can be directed to analysis programs such as Excel or Matlab, with the utility "simout" workspace. The general principles, which should be documented in your report, are listed below:

1. Model 1: The Mass-Spring
   For this and any model, the physical situation should be sketched to show all components and operations, such as depicted in Fig. A.4. Next, the equations of motion should be written, with initial assignment of parameters to the components. Test the predictive power of the model with various input parameters. You should use parameter vectors in gain blocks to test several parameters in a single run, as shown in class. You will need to measure amplitudes and frequencies of your output and compare them with theory. This is done most accurately by exporting your output to Matlab (Fig. A.5).

Spring-mass

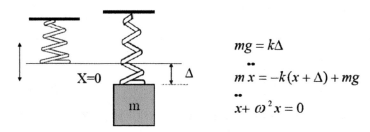

$$mg = k\Delta$$
$$m\ddot{x} = -k(x+\Delta) + mg$$
$$\ddot{x} + \omega^2 x = 0$$

Fig. A.5 Mass-spring model. When the unweighted spring, with stiffness, k, shown at left is attached to a mass, m, it is pulled down by the weight and lengthens by Δ, due to the force mg. When the spring is let go, and allowed to oscillate, it accelerates up and down, according to the dynamic equations. To find the frequency and amplitude of the oscillations, the equation of motion can be integrated by Simulink

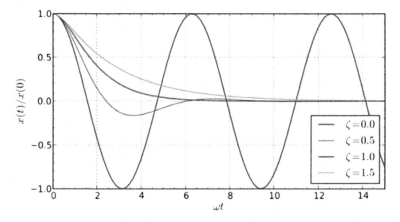

Fig. A.6 Mass-spring-damper behavior

2. Model 2: Mass-Spring-Damper
   For this, you add a damper to your spring model. The damper is a velocity-dependent component in parallel with the spring in Fig. A.4. The damping constant, c, can be formulated as a unitless parameter, by the following:
   Critical damping constant: $Cc = 2*\sqrt{(k*m)} = 2*m*w$; Unitless damping factor $\zeta = c/Cc$.
   Note how damping affects the output. You can test underdamping, overdamping, and critical damping (Fig. A.6).
3. Model 3. The Simple Pendulum
   Assumptions: The pendulum is a massless rod of length, L, to which is attached a mass, m, located at the center of mass of the pendulum. The model will be accurate when oscillations are small.

Fig. A.7 Free body diagram of passive pendulum

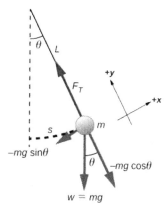

As seen in Fig. A.7, the pendulum experiences a torque, mg*L*sin(θ), that will cause it to swing clockwise, when it is released. Without resistance, the pendulum will swing to an angle, −θ, where it will experience a counter-clockwise torque, mg L sin(−θ). In this way, the pendulum swings back and forth between +θ and −θ, with torque always resisting the motion. In other words, the torque always opposes the motion.

Next the equation of motion is derived by applying basic trigonometry and Newton's second law. This law, when applied to rotational motion, states that the angular acceleration is $m*g/I$, where I is the moment of inertia = mL^2. Thus:

$$d^2(\theta/dt^2) = -mg*L*\sin(\theta)/I \text{ which reduces to}: d^2(\theta/dt^2)$$
$$= -gg - LL \quad \theta \text{ (for small angle approximation } \sin(\theta) \sim \theta)$$

where g is gravity, L is length of the pendulum, and θ is angle of the pendulum from zero (straight down) (Figure 6 Pendulum model).

Computing theta is done by double-integrating the acceleration. To start motion, we start with the initial angle of the pendulum in Fig. A.7, as the initial condition = 1 radian =~57°. When the pendulum is let go, our equation for theta is:

$$\Theta = 1 + \left(\int \int_0^{N\delta} \frac{d^2}{dt^2} \frac{-g}{L} \theta \ dt \right)$$

Here, theta is computed by double-integrating the acceleration. To start motion, we start with the initial angle of the pendulum in Fig. A.7, as the initial condition = 0.5 radians = ~30°. When the pendulum is let go, our equation for theta is as above.

> Here, the integrations are done at N intervals of time, δ. The inner integral calculates the velocity, $\dot{\theta}$, at each iteration. Then the outer integral calculates (t) at each iteration. A dimensional analysis of each term in the equation should be done to check its correctness. Since the units of (t) and its initial condition are both radians, the double integral must also be radians. So, dimensional analysis of the inner integral yields:

The second (outer) integral yields the angle, in units of radians, as expected. To implement a method to calculate Eq. (1), we can use integrator blocks found in Matlab-Simulink. Integration is done using the following method in Simulink.

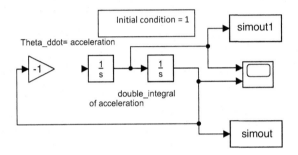

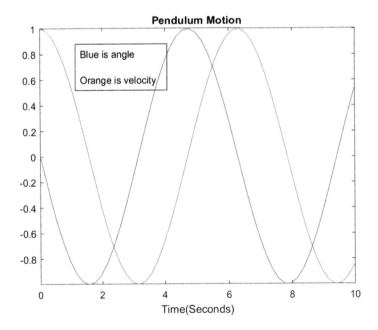

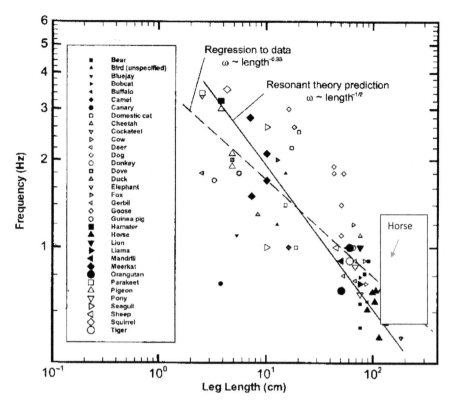

Fig. A.8 Relationship between leg length and the natural frequency of leg oscillation in a variety of animals

Here we see that the pendulum starts at +1 radian, where we set the initial condition in the second integral.

The pendulum swings from +1 to −1 in a sinusoidal pattern.

Velocity peaks at the bottom (0 radians) of each swing.

To test the validity of your model, you must check its output against the expected output. Looking at the equation of motion, we see that two parameters govern the motion: gravity and length. So varying either (or both) parameters should affect the output in a predictable manner. Assuming the model applies to a swinging leg, we can vary the length of the leg. As seen in Fig. A.8, there is an approximate log-linear relationship between frequency of oscillation and the leg length. This can be readily tested to verify your model.

4. Model 4. The Leg: A Pendulum with Resistance

A more realistic model of the straight leg swing can be represented as shown below. Here the springs can represent the resistance of the joint and tendons during swinging. Deriving the equations of motion can be done with a bit of trigonometry that you should try. A simplification is combining the two springs into a single spring with constant, k, since both are resisting the motion.

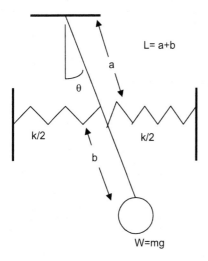

Hints About Simulink

The program is a digital-analog solver, and thus produces digitized outputs. Operation modes of the program can be modified by the various tabs at the top of the menu. For example, the sampling rate can be changed in the "Simulation" tab by selecting "model configuration parameters." The default configuration is "variable-step," with a tolerance of 1 e-3. This means that the sampling rate is adjusted to maintain error less than the tolerance. For certain applications you may want to change the tolerance. You may also want to use "fixed step" for a constant sampling

Appendix: Computational Modeling

rate. This selection may be desirable for seeing pendulum motion. Feel free to experiment with settings in order to understand them. You should also familiarize yourself with the various blocks.

Modeling Data with Matlab

Introduction

Raw biomechanical data contain information about movement, but inevitably are contaminated with noise from the sensors, measuring devices, and motions that are not associated with the human movement, i.e., external vibrations. Interpreting the movement data accurately thus requires processing the signal to remove extraneous signals, without corrupting the actual biomechanical data. While data filtering inevitably removes some parts of the desired signal, a fair representation of the actual movement can usually be obtained. Since human movements are highly variable, their kinematic data are further analyzed using modeling techniques, as described in this chapter. For example, Fig. A.9 shows a raw (filtered) velocity waveform obtained from a subject's arm during a flexion. Extracting. These may include filtering as well as modeling. Interpreting movements in different contexts requires comparing the actual movement against a theoretical model, as shown in Fig. A.9. The data can approximately represent the movement, but they are contaminated with noise. To accurately interpret the data, the measurement noise needs to be filtered out. Additionally, each human movement has idiosyncrasies that can complicate matters in certain applications, e.g., building a simple model of the movement of the elbow in a constrained movement task. For this, the observational (raw) data can be simplified by fitting analytical curves to them, as discussed herein.

Goal To learn basic curve fitting and build an understanding of when and how to implement a continuous versus a discrete optimization approach (Fig. A.9).

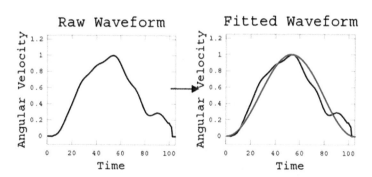

Fig. A.9 waveform (*Left*) and a 3-parameter fitted bell curve (*Right*). Abscissa represents percent of motion from full extension to full flexion

Overview: Here we will produce a MATLAB script that will fit a simplified model to raw kinematic data recorded from the lab: angular velocity of the elbow joint in planar motion. This process is an important aspect of biomechanical analysis and modeling and is the basis on which motor programs are created in robotics applications. The activities performed in this lesson can be performed in the basic version of MATLAB, but can be made more efficient with the Optimization Toolbox. The code will be written sufficiently general such that a wide variety of 1-dimensional data can be fit without substantially altering the code.

WARNING: The crux of this lesson is Optimization. Optimization calculations can be very intensive. The activities performed in this lesson should require less than 2 s to complete on a modern desktop, but it is possible for some of these exercises, including the homework, to take many seconds (10 or more). You can always break a calculation using **CTRL**

+ **C** at the command prompt.

Table of Contents

Section Title	Page
0. Getting Started	2
1. Basic curve fitting	5
2. *de novo* curve fitting	9
3. Matching over a discrete variable	12
4. Matching over a continuous variable	16
5. Summary	19

Getting Started

This exercise is designed to function as a stand-alone introduction to the repetition extraction problem, without pre-existing specialized skill in programming. All figures and text presented herein are based on the `five_cycles.txt` dataset, downloadable at Dextrahand.org.

To begin working with the dataset, save the file into the desired folder (for example, `C:\Classes\Measurements`) and open MATLAB. Create a new m-File and save that into the same folder (e.g., as `CurveMatching_Lesson.m`). The following is a snippet of code for reading in and plotting the data.

Appendix: Computational Modeling

CurveMatching_Lesson.m
```
1    close all; clear all;
2
3    %%% read in data from file, differentiate into velocity
4    posn_data=dlmread('fiveflexions_data.txt');
5    vel_data=diff(posn_data,[],2);
6
7    %%% set a few plotting parameters
8    label_size=20;
9    title_size=25;
10   tick_size=15;
11   line_width=4;
12
13   %%% plot the cyclic elbow flexion (velocity) data
14   figure
15   subplot(1,2,1)
16 set(gca,'position',[0.12 0.30 0.36 0.40])
17   plot(transpose(vel_data),'linewidth',line_width)
18   %%% make plot box pretty (turn off box, increase font size)
19   set(gca,'box','off')
20   set(gca,'fontsize',tick_size)
21   %%% expand plot window size to (min-2%) and (max+20%)
22   xlim([0 size(vel_data,2)+1])
23              plt_rng=max(vel_data(:))-min(vel_data(:));   23
ylim([min(vel_data(:))-(0.02*plt_rng),...
          max(vel_data(:))+(0.20*plt_rng)])
24   %%% annotate
25   xlabel('Time (sample number)','fontsize',label_size)
26   ylabel('Velocity (degs/sec)','fontsize',label_size)
27     title_string=sprintf('%s\n%s','Cyclic  Biomech-','anical Dataset');
28   title(title_string,'fontsize',title_size)
29

30
31 subplot(1,2,2)
32 set(gca,'position',[0.60 0.30 0.36 0.40])
33   plot(vel_data(1,:),'linewidth',line_width)
34   %%% make plot box pretty (turn off box, increase font size)
35   set(gca,'box','off')
36   set(gca,'fontsize',tick_size)
37   %%% expand plot window size to (min-2%) and (max+20%)
38   xlim([0 size(vel_data,2)+1])
39   plt_rng=max(vel_data(:))-min(vel_data(:));
40 ylim([min(vel_data(:))-(0.02*plt_rng),...
```

```
                   max(vel_data(:))+(0.20*plt_rng)])
41     %%% annotate
42     xlabel('Time (sample number)','fontsize',label_size)
43     title_string=sprintf('%s\n%s','Single','Data Cycle');
44     title(title_string,'fontsize',title_size)
45     %%% save figure
46     print('-dpng','F01_CyclicData.png','-r100')
47
48 A few notes on this code:
  5     vel_data=diff(posn_data,[],2);
```

The derivative is taken with the diff function. The diff operation takes the sample-to-sample differences within a column. But our data is oriented in rows (size(posn_data) shows that this matrix is five rows by 105 data points), so we specify that we want to take the difference *within a row*, or equivalently: across columns. The dimension of differentiation (here the column dimension, i.e., dimension #2) is specified as the third input argument to diff. So, since we accept the default method for taking a difference (so specified by placing an empty bracket [] in the place of the second input argument), we identify dimension number 2 in the third argument. If this is not clear, set x to be some small matrix, e.g., x=magic(4), and try differentiating in both directions: diff(x), and diff(x,[],2).

```
17 plot(transpose(vel_data),'linewidth',line_width)
```

As described above, the data are oriented with each movement cycle in a separate row. This runs against the grain of the MATLAB defaults, which presumes that time runs across rows, not columns. So plotting the velocity data straightaway via plot(vel_data) would produce a very strange result. We invoke the plotting command on a dataset transposed to conform to the default expectation of MATLAB's plotting routine.

```
   28    title_string=sprintf('%s\n%s','Cyclic   Biomech-','anical
Dataset');
```

MATLAB's plotting engine presumes by default that the title of a figure will be short enough to capture on a single line. This is usually sufficient, but for cases where the title is long (or, possibly: the font size is large and pushes the title into the x-label or off the page entirely), we can gather the text into a series of lines using sprintf. This function has two sets of arguments. The first is an amalgam of *metacharacters* %s (single string) and \n (new line). The subsequent arguments are the two single strings themselves.

```
34 plot(vel_data(1,:),'linewidth',line_width)
```

This may cause some confusion in the context of the above transposition: it is true that when plotted ensemble, MATLAB takes matrices and plots each column as

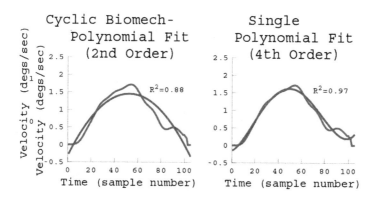

Fig. A.10 First glance at velocity data ensemble (*left*) and a single track (*right*)

a separate line plot (and since our data was arranged *across* columns, we transposed the matrix before plotting). When plotting a single dataset, however, i.e., when the object being plotted has either a row or column dimension of size 1, then it does not matter whether the object is a row or column vector: it will plot correctly. Here, we are plotting a row vector, which plots as we would like it to, because it is just a vector (and not a matrix). No need for the transpose for single vectors (Fig. A.10).

A programmer's first instinct should be to *look at* the data. Now that we have had a glimpse at the data, let us begin our analysis.

Basic Curve Fitting

There are quite a few ways to fit a curve to data. One of the most convenient and widely used approaches is that of the polynomial fit. A polynomial is a function on x of the general form

$$y = \sum^{n} k \cdot x_i \quad i = 0$$

where i iterates from 0 (coefficient) to n the polynomial order. For a second-order polynomial, this looks generally like

$$y = k_0 \cdot x^0 + k \cdot x^1 + k \cdot x^2$$

which is equivalent to

$$y = k_0 + k_1 \cdot x + k_2 \cdot x^2$$

Higher-order polynomials follow suit.

The basic shape of the velocity profile of a single-joint movement is abstracted as having a bell shape. The two simplest polynomials which adopt an approximately bell-like shape are polynomials of order 2 and 4. We set these up as follows:

CurveMatching_Lesson.m
```
>
>     %%% re-name vectors for convenience
>     y=vel_data(1,:);
>     x=1:length(y);
>     %%% create a polynomial fit
>     coeffs_2=polyfit(x,y,2);
>     yfit_2=(coeffs_2(1)*(x.^2)),...
            +coeffs_2(2)*x,...
            +coeffs_2(3);
>     coeffs_4=polyfit(x,y,4);
>     yfit_4=(coeffs_4(1)*(x.^4)),...
            +(coeffs_4(2)*(x.^3)),...
            +(coeffs_4(3)*(x.^2)),...
            +(coeffs_4(4)*(x.^1)),...
            +(coeffs_4(5)*(x.^0));
>
```

Note that we "short-handed" the phrasing on yfit_2, and wrote out each term explicitly for yfit_4. Either is acceptable; the coder should employ which ever convention is comfortable. Note further that we have arranged our code on multiple lines by using the , ... notation. This is a not just a typographical abridgement, but a legitimate coding practice: MATLAB recognizes these four characters as an indication that this line of code is to be continued on the next line. You may copy this code directly into MATLAB, and it will run correctly based on a verbatim paste.

We plot as:

CurveMatching_Lesson.m
```
>
>     %%% plot the cyclic elbow flexion (velocity) data
>     figure
>     subplot(1,2,1)
>     set(gca,'position',[0.12 0.30 0.36 0.40])
>     hold on
>     plot(x,y,'linewidth',line_width)
>     plot(x,yfit_2,'linewidth',line_width,'color',[1 0 0])
>     %%% imprint the goodness of fit as R^2
>     sse=sum((y-yfit_2).^2);
>     sse=sum((y-mean(y)).^2);
>     r_sq=1-(sse/sst);
>              text(length(y)*0.95,max(y)*0.95,['R^2=',num2str(r_sq,'%0.2f')],... 'fontsize',label_size-5,'horizontalalignment','right')
>     %%% make plot box pretty (turn off box, increase font size)
>     set(gca,'box','off')
```

Appendix: Computational Modeling

```
>     set(gca,'fontsize',tick_size)
>     %%% expand plot window size to (min-2%) and (max+20%)
>     xlim([0 length(y)+1])
>     ylim([-0.5 2.5])
>     %%% annotate
>     xlabel('Time (sample number)','fontsize',label_size)
>     ylabel('Velocity (degs/sec)','fontsize',label_size)
>   title_string=sprintf('%s\n%s','Polynomial Fit','2nd Order');
>     title(title_string,'fontsize',title_size)
>
>
>     subplot(1,2,2)
>     set(gca,'position',[0.60 0.30 0.36 0.40])
>     hold on
>     plot(x,y,'linewidth',line_width)
>     plot(x,yfit_4,'linewidth',line_width,'color',[1 0 0])
>     %%% imprint the goodness of fit as R^2
>
>     sse=sum((y-yfit_4).^2);
>     sse=sum((y-mean(y)).^2);
>     r_sq=1-(sse/sst);

>
text(length(y)*0.95,max(y)*0.95,['R^2=',num2st
r(r_sq,'%0.2f')],...
>     'fontsize',label_size-5,'horizontalalignment','right')
>     %%% make plot box pretty (turn off box, increase font size)
>     set(gca,'box','off')
>     set(gca,'fontsize',tick_size)
>      %%% expand plot window
>     xlim([0 length(y)+1])
>     ylim([-0.5 2.5])
>     %%% annotate
 size to (min-2%) and (max+20%)
  > xlabel('Time (sample number)','fontsize',label_size)
  > title_string=sprintf('%s\n%s','Polynomial Fit','4th Order');
  > title(title_string,'fontsize',title_size)
  > %%% save figure
  > print('-dpng','FO2_PolynomialFits.png','-r100')
>
>
```

With a few additional notes:

```
> hold on
```

This code allows multiple plots to exist on the same set of axes; without this code, each plot will replace the plot preceding it, and the axis will show only one plot at a time.

```
>
text(length(y)*0.95,max(y)*0.95,['R^2=',num2st
r(r_sq,'%0.2f')],...
```

The text command places an annotation on the axis. The documentation on this is very good, so we only make brief mention of its use here, but make special note of the third argument, i.e., the string for plotting. There are two tricks being employed here: string concatenation by square brackets, and truncation of a numerical quantity. String concatenation was achieved by [<string 1>, <string 2>], where string 1 in our case is 'R^2=' and string 2 is num2str(...), and num2str converts a numerical value (in this case, the r-squared calculation) into a string for display. While it is totally acceptable to invoke the default num2str(r_sq), we add an extra argument to cut the floating decimal (f) at two places to the right of the dot (0.2); the percent sign % merely flags this as a metacharacter. If this is not clear, practice with num2str(r_sq,'%0.3f') and num2str(r_sq,'%0.4f') at the command line.

The results are shown below (Fig. A.11).

Notice that increasing the polynomial order improves the fit, as shown by an R^2 near 1, in the case of the fourth-order polynomial. While this is a very convenient paradigm, there are compelling arguments against using polynomial fits in biomechanical measurements, the greatest being that they are thought to have little biological relevance: it is considered somewhat unlikely that the hand should move about the elbow in a way that is governed by a function that is polynomial in time. To evidence this, recall that we described the abstraction of the velocity profile as having a bell shape; neither of the fitted traces above have a bell shape: all even

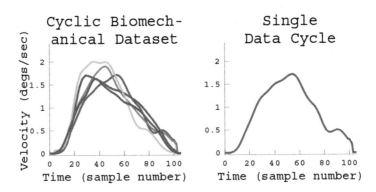

Fig. A.11 Second-order (*left*) and fourth-order (*right*) polynomial fits to velocity data

polynomials will continue forever toward infinity (or $-\infty$, as we are using negative polynomials); even if very slowly so.

De Novo Curve Fitting

Polynomial fitting is nice, but does not always get the job done. Sometimes, we have to "brute-force" our own curve fit. MATLAB has some additional functions for simulating curves, which are equally convenient. The `hanning` function does a nice job at simulating a bell curve and takes only a single argument: the length of the window. The Hanning window is a curve with a peak at $y = 1$ by convention, so we expect that we will have to change the amplitude in order to approximate our empirical data (it will not be well matched to our velocity which has a peak that is unlikely to match the peak of the Hanning function).

Note that the output of `hanning` is a column vector; we will transpose to a row vector for convenience...

```
CurveMatching_Lesson.m
>
>    %%% create the hanning window of the appropriate size
>    yfit_h=hanning(length(y));
>    %%% transpose to make row-vector (output was a col. vector)
>    yfit_h=transpose(yfit_h);
>
>    %%% plot the velocity trace and its hanning function
>    figure
>    subplot(1,2,1)
>    set(gca,'position',[0.12 0.30 0.36 0.40])
>    hold on
>    plot(x,y,'linewidth',line_width)
>    plot(x,yfit_h,'linewidth',line_width,'color',[1 0 0])
>    %%% imprint the goodness of fit as R^2
>    sse=sum((y-yfit_h).^2);
>    sse=sum((y-mean(y)).^2);
>    r_sq=1-(sse/sst);
>           text(length(y)*0.95,max(y)*0.95,['R^2=',num2str(r_sq,'%0.2f')],... 'fontsize',label_size-5,'horizontalalignment','right')
>    %%% make plot box pretty (turn off box, increase font size)
>    set(gca,'box','off')
>    set(gca,'fontsize',tick_size)
>    %%% expand plot window size to (min-2%) and (max+20%)
>    xlim([0 length(y)+1])
>    ylim([-0.5 2])
```

```
>      %%% annotate
>      xlabel('Time (sample number)','fontsize',label_size)
>      ylabel('Velocity (degs/sec)','fontsize',label_size)
>   title_string=sprintf('%s\n%s','Fit to a','Hanning Function');
>      title(title_string,'fontsize',title_size)

>
>
>      %%% normalize amplitudes (both traces have min of 0, so only
>      %%% need to div. by max in order to unity normalize)
>      y=y/max(y);
>      yfit_h= yfit_h/max(yfit_h);
>
>      subplot(1,2,2)
>      set(gca,'position',[0.60 0.30 0.36 0.40])
>      hold on
>      plot(x,y,'linewidth',line_width)
>      plot(x,yfit_4,'linewidth',line_width,'color',[1 0 0])
>      %%% imprint the goodness of fit as R^2
>      sse=sum((y-yfit_4).^2);
>      sse=sum((y-mean(y)).^2);
>      r_sq=1-(sse/sst);
>             text(length(y)*0.95,max(y)*0.95,['R^2=',num2str(r_
sq,'%0.2f')],... 'fontsize',label_size-5,'horizontalalignment','r
ight')
>      %%% make plot box pretty (turn off box, increase font size)
>      set(gca,'box','off')
>      set(gca,'fontsize',tick_size)
>      %%% expand plot window size to (min-2%) and (max+20%)
>      xlim([0 length(y)+1])
>      ylim([-0.15 1.15])
>      %%% annotate
>      xlabel('Time (sample number)','fontsize',label_size)
>             title_string=sprintf('%s\n%s','Hanning  Fit','Ampl.
Adjusted');
>      title(title_string,'fontsize',title_size)
>      %%% save figure
>      print('-dpng','FO3_HanningFits.png','-r100')
>
>
```

This is a straightforward curve matching with a single variable: duration of the Hanning window. The results are shown as follows (Fig. A.12):

Notice the vast improvement in goodness-of-fit with a simple parallel normalization (R^2 increase from 0.40 to 0.92).

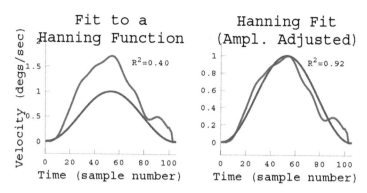

Fig. A.12 Hanning window fit to raw velocity (*left*) and unity-normalized velocity (*right*)

Matching over a Discrete Variable

The Hanning curve fit shown above illustrates the power of a well-chosen curve to model a biomechanical event: a single parameter was adequate to get very good similarity. Can we do better by adding another variable. The answer is almost always yes, and usually: this is a necessary activity. The sample velocity used here is rather well behaved; "real-world" data are rarely so easy to fit.

One avenue for adding a parameter is to change the width of the Hanning trace. The problem there is that the duration of our movement (at 105 points) is fixed. And any comparison trace would need to be length-matched. So we try a series of shorter traces (104, 103, ..., down to 5 points in duration), padding them with zeros as necessary. The basic approach is as follows:

We have already seen one Hanning function:

hanning (105)

But what if the data were better fit by a 104-point Hanning function? Then there are two possible 104-point vectors to try:

[*hanning* (104)   0]

and

[0   *hanning* (104)]

i.e., a shorter window padded either at the left or the right with another 0 (recall that a Hanning function is 0 at either end). Suppose that the data is best fit by an even shorter Hanning window, say of 103 points. In that case, there are

[*hanning* (103)   0   0],
[0   *hanning* (103)   0],
and   [0   0 *hanning* (103)]

A similar case could be drawn for a 102-point Hanning, a 101-point Hanning, and so on. We declare that a 5-point Hanning is the minimally sufficient Hanning function for us to be interested in modeling. While it is very unlikely that we would encounter a velocity trace that is best fit by a triangular wave form (the Hanning in 5 points) surrounded by 100 points of 0 distributed at the tails, modelers generally pride themselves on the exhaustiveness of their efforts.

We set up a for-loop as follows:

```
CurveMatching_Lesson.m
>
>       %%% set up data length (@N) and minimum bell width (@bw_min)
>       N=length(y);
>       bw_min=5;
>       %%% initialize dummy sum-sq-error value (arbitrarily high)
>       sse_val=inf;
>
>       %%% for bell curves between 5- and N-points long...
>       for i=bw_min:N
>       %%% for left-pads between 0- and (N-i)-points long...
>       for j=0:N-i
>       %%% set up the comparison waveform
>       bell_curve=transpose(hanning(i));
>       %%% set up the pads
>       left_pad=zeros(1,j);
>       right_pad=zeros(1,N-i-j);
>       comp_trace=cat(2,left_pad,bell_curve,right_pad);
>       %%% unity normalize
>       comp_trace=comp_trace/max(comp_trace);
>       %%% find sum of square error
>       temp_sse=sum((y-comp_trace).^2);
>       if temp_sse<sse_val
>       %%% retain fit variables
>       sse_val=temp_sse;
>       fit_trace=comp_trace;
>           fit_vars=[ length(left_pad),... length(bell_cuve),... length(right_pad) ];
>       end
>       end %%% end-for on j (left pad size)
>       end %%% end-for on i (bell curve size)
>
>       %%% plot the velocity trace and its hanning function
>       figure
>       set(gca,'position',[0.30 0.30 0.36 0.40])
>       hold on
```

Appendix: Computational Modeling

```
>     plot(x,y,'linewidth',line_width)
>     plot(x,fit_trace,'linewidth',line_width,'color',[1 0 0])
>     %%% imprint the goodness of fit as R^2
>     sse=sum((y-yfit_h).^2);
>     sse=sum((y-mean(y)).^2);
>     r_sq=1-(sse/sst);

>              text(length(y)*0.95,max(y)*0.95,['R^2=',num2str
(r_sq,'%0.2f')],... 'fontsize',label_size-5,'horizontalalignment',
'right')

>
text(length(y)*0.95,max(y)*0.85,
['L-Pad=',num2str(fit_vars(1),'%0.2f')],...
'fontsize',label_size-5,'horizontalalignment','right')

>
text(length(y)*0.95,max(y)*0.75,
['H-Width=',num2str(fit_vars(2),'%0.2f')],...
'fontsize',label_size-5,'horizontalalignment','right')

>
text(length(y)*0.95,max(y)*0.65,
['R-Pad=',num2str(fit_vars(3),'%0.2f')],...
'fontsize',label_size-5,'horizontalalignment','right')
>     %%% make plot box pretty (turn off box, increase font size)
>     set(gca,'box','off')
>     set(gca,'fontsize',tick_size)
>     %%% expand plot window size to (min-2%) and (max+20%)
>     xlim([0 length(y)+1])
>     ylim([-0.15 1.2])
>     %%% annotate
>     xlabel('Time (sample number)','fontsize',label_size)
>     ylabel('Velocity (degs/sec)','fontsize',label_size)
>              title_string=sprintf('%s\n%s','Hanning
Fit','(2-parameters)');
>     title(title_string,'fontsize',title_size)
>     %%% save figure
>     print('-dpng','FO4_HanningLoop1.png','-r100')
>
>
```

which is plotted as (Fig. A.13)

Fig. A.13 Plot of best-fit 2-parameter Hanning

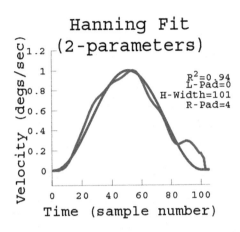

We make a few notes here:

>   sse_val=inf;

sets an initial value for the sum-of-squared error. Any trace will match better to y than an infinitely large error, so this will get overwritten immediately. But it must be initialized to some value, so that we can enter the `if temp_sse<sse_val` logic on the first pass: if the `sse_val` is never established, then the logical test will yield an error.

>

`comp_trace=cat(2,left_pad,bell_curve,right_pad);`

is a straightforward concatenation of three different items as row vectors (concatenated as rows by `cat(2`. The left pad, the bell curve, and the right pad are concatenated to create a single comparison trace. Note that there are three pieces to this trace, but only two parameters in the model. This is a "Degree of Freedom" phenomenon: We are constrained to a `comp_trace` of 105 points in length. Given that restriction, once the bell curve length is defined (by `i`), and the left pad length is defined (by `j`), the right pad length is given by simple subtraction. This is, correctly, a 2-parameter model.

Notice, lastly, that the best-fit Hanning trace (in a 2-parameter sense, with the constraint of exact amplitude match) is a 101-point window, with a 0-point left pad (and 4-point right pad).

Matching over a Continuous Variable

There is a family of optimization functions available in the MATLAB programming environment for finding the best parameter in a continuous variable, e.g., amplitude. The basic paradigm is as follows:

1. Construct a function to be optimized
2. Where possible, identify a reasonable guess for the neighborhood in which the optimum value might be found

Appendix: Computational Modeling

3. Enter function and the guess into one of the functions in MATLAB's Optimization Toolbox, e.g., fminsearch or fminunc

The only aspect of this routine that might be unfamiliar is the function statement (part 1), but this should at worst be a foreign concept; it is not a particularly difficult task. Here, we want to free ourselves of the presumption that the amplitude of the Hanning curve needs to be equal to the amplitude of our raw velocity.

```
CurveMatching_Lesson.m
>
>     %%% for bell curves between 5- and N-points long...
>     for i=bw_min:N
>     %%% for left-pads between 0- and (N-i)-points long...
>     for j=0:N-i
>     %%% set up the comparison waveform
>     bell_curve=transpose(hanning(i));
>     %%% set up the pads
>     left_pad=zeros(1,j);
>     right_pad=zeros(1,N-i-j);
>     comp_trace=cat(2,left_pad,bell_curve,right_pad);
>     %%% unity normalize
>     comp_trace=comp_trace/max(comp_trace);
>     %%% create a difference function for minimization
>     diff_fn=@(x) sum(((x*comp_trace)-y).^2);
>     %%% perform search for the unconstrained minimum
>     temp_amp=fminunc(diff_fn,1);
>     %%% find sum of square error
>     temp_sse=sum((y-(temp_amp*comp_trace)).^2);
>     if temp_sse<sse_val
>     %%% retain fit variables
>     sse_val=temp_sse;
>     fit_trace=comp_trace;
>         fit_vars=[ length(left_pad),... length(bell_cuve),...
length(right_pad),... temp_amp ];

>     end
>     end %%% end-for on j (left pad size)
>     end %%% end-for on i (bell curve size)
>
>     %%% plot the velocity trace and its hanning function
>     figure
>     set(gca,'position',[0.30 0.30 0.36 0.40])
>     hold on
```

```
>      plot(x,y,'linewidth',line_width)
>      plot(x,fit_trace,'linewidth',line_width,'color',[1 0 0])
>      %%% imprint the goodness of fit as R^2
>      sse=sum((y-yfit_h).^2);
>      sse=sum((y-mean(y)).^2);
>      r_sq=1-(sse/sst);
>              text(length(y)*0.95,max(y)*0.95,['R^2=',num2str(r_sq,'%0.2f')],... 'fontsize',label_size-5,'horizontalalignment','right')

> text(length(y)*0.95,max(y)*0.85,['L-Pad=',num2str(fit_vars(1),'%0.2f')],... 'fontsize',label_size-5,'horizontalalignment','right')

> text(length(y)*0.95,max(y)*0.75,['H-Width=',num2str(fit_vars(2),'%0.2f')],... 'fontsize',label_size-5,'horizontalalignment','right')

> text(length(y)*0.95,max(y)*0.65,['R-Pad=',num2str(fit_vars(3),'%0.2f')],... 'fontsize',label_size-5,'horizontalalignment','right')

> text(length(y)*0.95,max(y)*0.55,['Amp=',num2str(fit_vars(4),'%0.2f')],... 'fontsize',label_size-5,'horizontalalignment','right')
>      %%% make plot box pretty (turn off box, increase font size)
>      set(gca,'box','off')
>      set(gca,'fontsize',tick_size)
>      %%% expand plot window size to (min-2%) and (max+20%)
>      xlim([0 length(y)+1])
>      ylim([-0.15 1.2])
>      %%% annotate
>      xlabel('Time (sample number)','fontsize',label_size)
>      ylabel('Velocity (degs/sec)','fontsize',label_size)
>              title_string=sprintf('%s\n%s','Hanning Fit','(3-parameters)');
>      title(title_string,'fontsize',title_size)
>      %%% save figure
>      print('-dpng','FO5_HanningLoop2.png','-r100')
>
>
```

Fig. A.14 Plot of best-fit 3-parameter Hanning

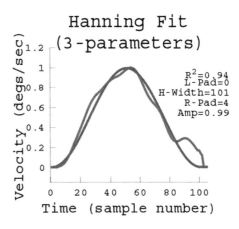

With the relevant changes to the code highlighted. Notice that the code has a very similar structure to the 2-parameter routine, with the major difference being the addition of the `fminunc` call. We plot these as (Fig. A.14):

We note here that the addition of a third parameter did not manifestly improve the fit.

This is a perfectly legitimate result.

Summary

In this lesson, we learned some basic principles of matching a simplified or idealized waveform to a raw data trace. The techniques put forth here only scratch the surface of what is possible in curve matching, but provide a solid foundation for implementation in an actual laboratory setting, and expansion via more complex waveforms.

Commands used here:

```
-close all,    clear all
-dlmread
-figure, plot, print, xlabel, ylabel, title, subplot, text
-set(gca,...   box, fontsize, position
-xlim, ylim
-length, size
-max, min
-find
-diff
-transpose
-sprintf, num2str (and metacharacters, via '%0.2f')
```

```
-hanning
-zeros
-cat
-inf
-for, if
-fminunc
```

Concepts employed here:

Default dimensions, and optional arguments to change defaults (e.g., `vel_data=diff(posn_data,[],2)`) Polynomials

Goodness of fit

Window functions (e.g., `hanning`) Normalization

Padding (e.g., by adding zeros)

For-Loops and Comparative Logic (via `for` and `if`)

Function declarations (e.g., `diff_fn=@(x) sum(((x*comp_trace)-y).^2)`) Optimization

Naturally, we saw code that was well formatted and adequately commented, and plots that were well annotated.

$$\tau_i = -k_i\left(\theta_i - \theta_{ei}\right) - b\dot{\theta}$$

$$\omega_0 = \sqrt{\frac{g}{L}}$$

Index

A
Above-elbow (AE) amputation, 13
Above-elbow (AE) loss, 85
Above-knee (AK) amputation, 12, 58
    artificial knees, 58
    gait cycle, 59
    lower limb prostheses, 58
Activities of daily life (ADL), 36
Activities of daily living (ADL), 170, 175, 190
Adaptive neuromotor learning, 112, 113
Adduction/abduction, 72
Advanced prosthetic knees, 60
Agonist/antagonist forces, 80
Ambulation, 9
Ambulatory efficiency, 160
Amputations, 9, 10
Anatomical bony landmarks, 32
Anatomical coordinates
    3-axis coordinate system, 36
    biomechanics, 37
    kinematic variables, 37
Anatomical terminology, 21
Ankle-knee interaction, 48
Ankle plantar-flexion, 41
Ankle prostheses, 42
Anterior insula, 114, 179
Anthromophometry
    biomechanical analyses, 33
    body segments, 32
    human body, 34
    human limbs, 33
    pendulum model, 35
    simplified coordinate system, 34
Arm actuators, 80
Arm major joints, 72

Arm motions dynamics and kinematics, 122
Artificial control system, 117
Artificial energy sources, 156
Artificial limbs, 2, 3
Axial loads, 18, 27

B
Base of support (BoS), 137
Beam analysis, 26
Beams buckle, 27
Below-elbow (BE) amputations, 13
Below-elbow (BE) socket, 135
Below knee (BK), 135, 136
Below-knee (BK) amputation, 10
    knee joint function, 52
    patellar ligament, 53
Biaxial stress equation, 29
Biceps, 18, 21
Biomechanics, 17
Biomimetic control, 107, 108
Biomimetic HMI, 107
Bion, 106
Block diagrams, 192
    feedback, 195
    input and controller output, 193
    leg oscillation, 199
    mathematical modeling, 194
    pendulum, 196
    physical system, 195
    reflex center, 193
    theta, 197
Body segments
    contours and volume, 36
    socket, 36

Body segments (*cont.*)
  volumes, 35
  water levels, 35
Bones distal, 75
Brain-machine interfaces (BMIs), 97, 98
Brain scanning, 179
Breathing effect, 111

C
Center of mass (COM), 126, 127
Cineplasty, 87, 88
Civil War, 3
Closed-loop configuration, 123
Closed-loop control logic, 119
Closed-loop feedback logic, 124–126
Closed-loop operation, 117
Congenital deformity, 8
Congenital limb deficits, 8
Control system, 117, 118
Cost of transport (COT), 156
Customized Training to Use Motor Prosthetic Systems, 177

D
Degrees of freedom (DoFs), 71–73, 77, 113
Desired position, 119
Dextra hand, 99, 107
Digital-analog solver, 200
Dynamics analysis, 23
Dysvascular foot lesion, 8
Dysvascularity, 1, 7

E
Effective mechanical advantage (EMA), 41
Elastic modulus, 18, 26
Elbow-disarticulation cases, 13
Elbow torque, 80
Electromyography (EMG), 44
  amplitudes, 93
  artificial hands, 113
  electrodes, 90
  signal, 89
Electro-rheological fluid (ERF), 64
Embedded Myokinetic Prosthetic Hand Controller, 93
Emerging technologies, 15
Energetics, ambulation
  artificial energy sources, 156
  COT, 156
  efficiency, 160
  energetic contributors, 158, 159
  energy conservation, 164
  energy fluctuations, 164
  gait restoration, 155
  lower-limb prostheses, 155
  measuring energy, gait, 157, 158
  mechanical-energy conservation, 164
  natural energy sources, 155
  oxygen consumption, 165
  power, 161
  powering electronic components, 155
  RSPs, 165, 166
  running-specific prostheses, 165
  testing, 167
  turtles, 164
  walking, 161–163
  work units, 161
Energy harvesting, 156, 162, 163
Equilibrium, 20
External assistive devices (EADs), 183
Externally powered prostheses (EPP), 85, 88
Extrafusal muscles, 128

F
Feedback-controlled gait system, 126
Feedback control system, 120
Femur, 11
Finite state machine, 59
Flexion, 53
Flexion-extension, 52
Flexor carpi ulnaris (FCU), 103
Flexor digitorum superficialis (FDS), 103
FMG-based HMI, 102, 103
FMG sleeve, 109
FMG transducer, 106
Force myography (FMG), 99, 110
Forearm rotational design, 80
Fourier-domain representation, 109
Fourth-order polynomial, 208
Free body diagram (FBD), 19, 191
  reaction force, 20
  rules, 19
  static situation, 21

G
Gait cycle, 56, 61
Gait kinetics, 44
Gait speed, 160
Gastrocnemius lateralis (cGL), 127
Gastrocnemius medialis (cGM), 127
Gastrocnemius muscles, 126
Gaussian noise, 44
Gravity, 155

Index 221

Grip force dynamometer (GFD), 110
Ground reaction forces (GRFs), 137, 138, 157

H
Hand, 72
Hand functionality, 72
Hand movement types, 78
Hanning curve, 211
Hanning function, 211
Hanning window, 209–211
Heating, ventilation, and air conditioner (HVAC), 117
Hemipelvectomy, 11
Higher-order polynomials, 205
Hip disarticulation, 11, 67
9-Hole peg test, 112
Human dexterity, 118
Human limbs, 22
Human locomotion, 46
Human-machine interface (HMI), 71, 178–180
  accurate sensing, 87
  adaptive neuromotor learning, 112, 113
  alternatives, 98, 99
  biomimetic control, 107, 108
  cosmetic restoration and agency, 114
  direct brain control, 97, 98
  EPPs, 85, 88
  FMG-based, 102, 103
  FMG *vs.* EMG, 99–101
  functions, 86
  increasing DoFs, 89
  IPPs, 85, 87
  local feedback, 111, 112
  MYOE characteristics, 91, 92
  MYOE signals, 89, 90, 93, 95–97
  prosthetic developments, 110, 111
  radio-transmitting electrodes implants, 92
  sensory receptors, 111, 112
  silicon sleeve, 104
  targeted muscle reinnervation, 92, 93
  volitional signal processing, 108, 109
Human movements, 25
  joint motions, 37
  kinematic variables, 37
Human-operated manipulandums, 15
Human walking kinetics
  anthropomorphic data, 38
  axial force, 39
  computational methods, 40
  and dynamics, 39
  motion analysis, 40
  whole-body angular momentum, 38
Humerus, 13
HVAC control system, 118

I
Immediate postoperative prosthetic (IPOP) procedure, 14
Implicit motor control (IMC), 182
Implicit motor control training (IMCT), 112
Inertial measurement units (IMUs), 38
Instantaneous COT (ICOT, 162
Insula activity, 114
Internally powered prostheses (IPP), 85, 87
International Society for Biomechanics, 34
Intrafusal muscles, 128
Inverse dynamics, 24, 25

J
Jaipur knee, 59
Joint dynamics
  dorsiflexor muscles, 40
  plantarflexion moment, 41
  torques, 40

K
Kinematics, 17, 22
  analyses, 38
  behavior, 22
  and dynamics, 23
Kinetics, 17
Knee disarticulation, 11
Knee flexion, 48
Knee jerk reflex, 128, 129, 194

L
Laminates, 146, 147
Laplacian operator, 109
Learning capability, 113
Leg functional anatomy
  Knee joints, 52–54
  muscles, 54
  quadriceps, 52
Ligatures, 2
Limb function, 31, 173
Limb-prosthetic interface
  BK residuum, 135, 136
  custom designing, 150, 151
  dynamic strengths, 149
  endoskeletal, 152, 153
  exoskeletal interface device, 133

Limb-prosthetic interface (cont.)
  fitting limb, 150, 151
  forces, 143
  GRFs, 137, 138
  laminates, 146, 147
  leg prosthesis, 139–141
  LL socket fitting, 134
  materials
    properties, 147, 148
    selection, 146
    sockets, 145
  optimal performance, 133
  protecting tissue, 145, 146
  residuum, 149, 150
  socket environment, 133, 134
  socket-skin interface, 142, 143
  structural connections, 133
  UL socket, 134, 135
  user, 133
Linearization equation, 109
LL socket fitting, 134
Locomotion, 9
  gait analysis, 46
  initial contact, 48
  kinematics and dynamics, 45
  knee flexion, 49
  lower limbs, 45
  mid-swing, 49
  push-off phase, 48
  step and stride lengths, 47
  swing phase, 49
  terminal deceleration, 49
  walking subjects, 46
Lower-extremity prostheses, 111
Lower-limb prosthetic developments, 62

M
Magneto-rheo knee, 65
Magneto-rheological Fluid (MRF) knee, 64
Man-machine systems, 109
Mass-spring, 195
Mass-spring-damper, 196
Mathematical modeling, 190, 194
MATLAB
  biomechanical data, 201
  dataset, 202
  diff operation, 204
  kinematic data, 201
  matrices and plots, 204
  movement data, 201
  programming environment, 214
  script, 202
  theoretical model, 201
  velocity data, 205

MATLAB-Simulink, 198
Mechanical testing, 27
Mechanical vibration-based energy harvesting systems, 163
Medio-lateral ankle moments, 42
Metacarpophalangeal (MCP), 76
Metallic prostheses, 2
Mirror Therapy, 177
Modeling process, 190
Modern imaging and materials technology, 51
Modular Limb Prosthesis (MLP), 169
Monticuli, 77
Motor control activation, 89
Motorized knee, 63
MRI, 98
Multimode fiber (MMF), 106
Muscle spindles, 120
Muscular activity, 49
Musculo-skeletal system, 190
Muybridge strobes, 38
Myoelectric (MYOE) signals, 88, 89
Myokinetic, 92
Myoplasty, 6
Myo-pneumatic (M-P), 98

N
NASA grip force-multiplying *Roboglove*, 85, 86
National Academy of Sciences, 4
Natural limb motion control system
  arm manipulation, 120
  arm movements, 120
  closed-loop control, 119
  components, 118, 120, 121
  dynamics and kinematics, 118
  feedback control system, 120
  kinetic and dynamic parameters, 119
  primary variables, 118
Naturally conditioned communication channels (NCCC), 113
Near-infrared signals (NIRS), 104
Negative feedback system, 128, 129
Neural prosthetic proprioception, 106
Neuromotor system (NMS)
  adaptability, 122–124
  artificial control, 117, 118
  closed-loop feedback, 117, 124–126
  controlled falling, 126
  desirable trajectory, 119
  gravity, 124, 126
  limb motions, 119
  limbs natural control, 118–122
  positional error monitoring, 119
  prosthetic control, 122

Index

tactile feedback, 129, 130
virtually infinite patterns, 117
walking coordination, 126, 128, 129
Newton's laws, 192
Newtonian mechanics, 17
Normal feedback channels, 112

O

Open and closed loop control, 193
Open-loop control, 117
Open-loop system, 118
Operant-conditioned communication channels (OCCC), 113
Optimal prosthetic combination (OPC, 183
Optimal trajectory, 119
Optimization Toolbox, 202
Optimum speeds, 161
Orthotics and prosthetics systems, 112
Osteomyelitis, 8

P

Palm and digits, 72
Patella, 52
Patellar-tendon-bearing (PTB), 135
Patella-tendon bearing (PTB) sockets, 144
Pendulum model, 197
Peripheral Nervous System-Machine Interface (PNS-MI), 93, 180
Phantom limb (PL), 114, 179
Physiological control system, 193
Piezoelectricity, 163
Pinching and palmar prehension, 74
Pneumatic FMG sensors, 101
Polynomial, 205
Polynomial fitting, 209
Postsurgical period, 14
Postures, 80
Potential energy, 159, 160
Prehension, 5, 78, 79, 81
Prosthetic advancements, 7
Prosthetic advances
    artificial limb, 3
    LL and UL, 6
    prosthetic socket, 4
Prosthetic and orthotic (P&O) devices, 15
Prosthetic ankles, 42
Prosthetic arm, 5
Prosthetic controller, 122
Prosthetic designs, 4
    agency
        characteristics, 181

control software, 183
Dextra and BioPatRec, 183
HMI-E sensor array, 182
muscle mappings, 182
nondominant limb, 182
powered control, 182
prescriptions, 183
quantitative performance measure, 183
restoration outcome metrics, 183
STP scheme, 183
STP system, 182
treatment combination optimization, 184
user, 181
virtual prosthetic equipment and environment, 182
    dexterity, 170–172
    direct brain control, 173
    exploiting adaptability, 181
    hand function, 176–180
    improving manipulation, 170
    restoring agency, 180
    sensory feedback, 170–172
    technology, 169
Prosthetic developments, 110, 111
Prosthetic knees, 56
Prosthetic motors, 112
Prosthetic/orthotic arm manipulation, 120
Prosthetics market, 15
Proximal interphalangeal (PIP), 76
Pseudosensory feedback signals, 112

Q

Quadriceps, 52, 53

R

Raw biomechanical data, 201
Referred phantom sensations, 114, 179
Representative specimen, 25
Residual kinetic image, 105
Residuum, 51
Restorative surgical procedures, 2
Restoring agency, 180
Restoring ambulation
    BK amputees, 68
    hip during swing phase, 68
    rhythmic gait, 67
Restoring limbs, 1
Revolutionizing Upper Limb Prosthetics, 169
Robotic arm, 112
Running-specific (RSP) prostheses, 165

S

Selfness, 114
SEMG records, 99, 100
Sensations, 130
Sensory feedback, 114
Sensory substitution, 171
Shear, 143
Shearing forces, 18
Shoulder disarticulation, 13
Signal response matrix, 103
Silicon sleeve with integrated FSR sensors, 104
Simulate, Test and Practice (STP), 182
Simulation, 177
Simulink, 189, 195
Single-joint movement, 205
Socket, 31
Socket design, 111
Socket environment, 133, 134
Socket-skin interface, 142, 143
Sophisticated upper-extremity systems, 111
Standard muscle stretch reflex, 128
Stimulating devices, 129
Straightforward curve matching, 210
Stress-strain curve, 25
Stress-strain relationship, 28
Stretch reflex sensors, 121
Sulci, 72
Surface EMGs (sEMGs), 44
Surface muscle pressure (SMP), 99
Surgical techniques
　limb loss, 7
　medical and surgical treatment, 7
　myoplasty, 6
　residual limb, 6
Synergy, 82

T

Tactile and kinesthetic sensitivity, 78
Tactile flow, 171
Terminal deceleration, 49
Thumb, 74
Tibial-Femoral Joint (TFJ), 53
Torques, 18

U

U.S. Civil War, 3
UL joint torques, 81
UL prosthesis, 98
UL prosthetic design, 81
　above-elbow (AE) amputation, 175
　ADL, 175
　arm prostheses, 174
　artificial human limb, 175
　clinical application, 175
　unilateral below-elbow amputees, 175
　user ranking of utility, 175
　user responses, 174
　user survey, 175
Ulna/radius rotation (UR), 103
Unconstrained isotropic cube, 27
Upper-extremity amputations
　devices and techniques, 13
　partial hand amputations, 12
　restoring, 12
　shoulder disarticulation, 13
　wrist disarticulation, 13
Upper limb (UL)
　amputees, 85
　anatomical positions, 74
　arm actuators, 80
　ball-and-socket joint, 72
　functional restoration (*see* Human-machine interface (HMI))
　hand and wrist motions, 76
　hand control, 78
　hand function, 72
　hand synergies, 82
　hand/wrist positions and forces, 73, 74, 77
　HMI, 71
　joint torques, 81
　joints axes, 72, 73
　manipulations, 80
　muscle groups, 71
　prosthesis, 5
　restoring functions, 71
　sockets, 134, 135

V

Vibrotactile feedback, 172
Virtual reality (VR), 177
Volitional signal processing, 108–110

W

Walking coordination, 126, 128, 129
Walking energy harvesting, 162, 163
Weight-bearing forces, 11
Wrist flexion and extension forces, 78

Y

Young's modulus, 26